Custom Search – Discover *more:*

Custom Search – Discover *more:*

A Complete Guide to Google Programmable Search Engines

Irina Shamaeva and David Galley

CRC Press
Taylor & Francis Group
Boca Raton London New York

CRC Press is an imprint of the
Taylor & Francis Group, an **informa** business
A CHAPMAN & HALL BOOK

First Edition published 2021
by CRC Press
6000 Broken Sound Parkway NW, Suite 300, Boca Raton, FL 33487-2742

and by CRC Press
2 Park Square, Milton Park, Abingdon, Oxon, OX14 4RN

Library of Congress Cataloging-in-Publication Data
Names: Shamaeva, Irina, author. | Galley, David, author.
Title: Custom search - discover more: a complete guide to Google programmable search engines / Irina Shamaeva, David Galley.
Other titles: Google.
Description: First edition. | Boca Raton : CRC Press, 2021. | Includes bibliographical references and index.
Identifiers: LCCN 2020048129 | ISBN 9780367567392 (paperback) | ISBN 9780367569686 (hardback) | ISBN 9781003100133 (ebook)
Subjects: LCSH: Search engines.
Classification: LCC TK5105.885.G66 S53 2021 | DDC 025.04252–dc23
LC record available at https://lccn.loc.gov/2020048129

ISBN: 978-0-367-56968-6 (hbk)
ISBN: 978-0-367-56739-2 (pbk)
ISBN: 978-1-003-10013-3 (ebk)

Typeset in Minion
by SPi Global, India

Contents

Introduction

GOOGLE'S CUSTOM SEARCH ENGINES (CSEs) offer search opportunities that are unavailable with any other tool. CSEs have advanced settings and search operators that do not work in regular Google searches. Anyone can create CSEs for themselves and others to use, uncover extra results, and boost research productivity.

With this book, we hope to popularize this fun and powerful tool so that many more people become aware of it and include CSEs in their research.

To our disappointment, Google has recently renamed "Custom Search Engines" to "Programmable Engines." But we like the old, long-lived name better and will continue using it.

You might be wondering why the word *more:* in the title is not capitalized and followed by a colon. It is intentional and has a double meaning (!). You will find out what we meant in the second half of the book.

We want to thank several colleagues who generously spent their time reviewing the manuscript draft: Julia Tverskaya, Elena Pavlovskaia, Pierre-André Fortin, Glenn Gutmacher, and Guillaume Alexandre. Our sincere thanks for your honest, constructive, and encouraging feedback.

GOOGLE vs. CSEs

"Google it!" Most people search Google with a few keywords looking to find one piece of information and find what they need as a first or second result.

As a researcher, however, you need to get some control over results and are often looking for as many results as possible, not just one. As a practical example, you might be looking to find social site profiles that fit additional requirements, such as a location, employer, or job title. Or you might be interested in top management moves in manufacturing. You can find the target information on Google, but you would need to run a series of advanced searches that are unlike "simple searching with keywords."

Google allows you to control results with advanced search operators, advanced search dialog, and settings. You can narrow the search to a given site, file format, or words in the page title or URL. For example, you can search for pages on the site slideshare.net with the word "CV" in the title. But you cannot filter results by the type of content pages represent, such as people or company profiles.

CSEs are a software layer Google built on top of its search. CSEs get results by filtering search results from Google to match the CSE settings. They allow you to use the same operators and settings as Google's. They also have additional parameters and search operators.

CSEs can do many things that Google cannot. You can narrow down to a country or language. You can set "soft" parameters that would make your CSE rank pages higher if they have given keywords or are on a given site. You can run long OR searches, overcoming Google's 32 keywords limitation.

What is of particular interest to us is that CSEs can search for standard types of structured information and values. You can search for people, restricting results to a given employer, job title, location, or school. You can search for *Microsoft* as an employer or *director* as a job title, not just a keyword. You can achieve the filtered search writing out advanced CSE operators. We will present detailed explanations of the operators' syntax, along with examples for you to recreate.

THE GAP TO COVER

Lots of professionals use existing CSEs, for instance, in talent acquisition to source for "passive" job candidates, in digital journalism or with other research goals. But few people have experience creating them. Even fewer know about advanced CSE-only search operators.

We think that the main reason CSEs are not as widespread as they should be is that it is not easy to get educated on CSE creation There is little information online and no books (other than the one you are reading). Only sketchy information is available on the "structured" operators, which are by far, the best CSE feature.

Another reason for the lack of popularity of CSEs is that Google's help and posts are oriented towards website owners. What is of interest to us are CSEs that search on sites that we did not build.

Help documentation is also not aligned 1-to-1 with what you see. Many links in Google's CSE support documentation refer to guides for programmers. It is as if they do not believe anyone will create or edit CSEs via the regular user interface (UI, called the Control Panel).

Yet another reason is that the terminology is too technical, and the UI is archaic. Context, Annotations, Refinements, Knowledge Graph, and Schema.org Objects on the front "basic" page in the Control Panel, as well as "JSON API," do not communicate anything to non-coders. Beginners get discouraged.

You can find some educational videos about CSEs, but none of them go in-depth.

We wrote the book to cover the gap.

CUSTOM SEARCH ENGINE USERS

There are two ways to engage with CSEs: as a creator and as an end-user.

- The creator can set up CSEs for themself and others (e.g., less technically savvy colleagues or clients). Even simple CSEs can be quite useful. But to take advantage of more advanced functionality, the creator needs a deeper understanding of how Google search and CSEs work.

- CSEs allow end-users to perform advanced searches without the need to use (or even understand) complex operators. For the more sophisticated end-user, CSEs' unique search operators allow for a high-precision web search unavailable anywhere else.

If you want to become an advanced CSE end-user and creator and search like Miss Marple, this book is for you. We highly recommend following each link and recreating each CSE we describe.

WHAT WE SKIPPED

In the book, we focus on creating and using CSEs to enhance web searches and explain features that Google does not cover in its documentation.

We do not expect readers to have a scripting background or marketing or site-building goals. Consequently, we have chosen not to cover specific features, such as usage statistics, advertising, and promotion, customizing the end-user UI, or programmatic implementations of CSEs.

PRIVACY AND DISCLAIMER

In cases where searches yield results showing non-public individuals, we have blurred or masked individually identifiable information out of respect for their privacy. This instructional book is not affiliated with nor endorsed by Google.

ERESOURCES

This book has been formatted with embedded hyperlinks to completed searches to facilitate access for eBook users. Readers of the printed versions of this book are encouraged to download their free eResources, including a full listing of these links as well as other published website links, from the Publisher's website at https://www.routledge.com/9780367569686.

1

Introduction to Google Custom Search Engines (CSEs)

CHAPTER 1

Simple and Advanced Google Search

B ACK IN 2007, OUR agency <u>Brain Gain Recruiting</u> placed a software engineer we found on Google. A few minutes' effort (plus, of course, all the interactions) made us about $25,000. We were quite impressed. We also enjoyed what felt like "treasure hunting" on Google. It was both educational and fun.

Since then, we have been digging deep into Google and using creative search strings to get hard-to-find results.

If you are in the business of finding information, advanced Google is your best friend. Custom Search Engines are an elaborate software layer on top of Google that you can utilize as well.

To understand how CSEs work, you first need to get familiar with Google search and its advanced operators. We have included a section about Google.com search and the advantages of using CSEs vs. searching on Google. They should provide you with some background, but You must run and modify the examples we have provided.

Let us start with two definitions related to Google search.

- *Search engine results pages* (SERP) are web pages served to users searching for something with a search engine, such as Google. The user enters their search query (often using specific terms and phrases known as keywords), upon which the search engine presents them with a SERP (*source*: https://www.wordstream.com/serp).

- *Search result snippets* are the additional context included with each result on the search results page. Google displays them under clickable links to resulting web pages, which it shows along with the page titles. Snippets include blocks from the page text highlighting your search terms in bold font. They may also show specific-to-the-webpage attributes, such as recipe ratings or event dates. They may show the last updated dates for the page. They are an essential tool to help searchers to find what they are looking for. (We picked part of the explanation from this source.)

SIMPLE GOOGLE SEARCH

"Google it!" Most people Google simply by entering a few keywords and look to find one "right" piece of information. Google search is optimized to provide one answer to your question. Users nearly always see the desired result as the first or second on the search results page.

In recent years, Google seems increasingly focused on providing a single "best" result for any search. In addition to standard search results, Google shows Knowledge Graph entities (to the right of search results) and Featured Snippets. Featured Snippets are an expanded preview of information from a site Google considers authoritative. They appear above the first search result. We will cover Knowledge Graph entities in more depth later in this book.

Here are some examples of Google providing answers to clear questions. We have provided multiple examples to show the broad scope of queries for which Google would give you "the" answer along with the search results. Click the links to view them. (As time goes, Google's ranking algorithm will get better in semantic search because it is a learning algorithm.)

- what is DNA

- how many people live in New York

- Paris to Hamburg

- COVID-19 statistics

- places to see in Croatia

- baby wipes

- Walgreens near me

- how far is the moon

- what is the tastiest Russian food

- how many people work at Google

- what is the difference between Java and javascript

- who is the president of Epam

- who is CTO at Oracle

- where are Google's offices in Europe

- how many people work at Deutsche Bank

- what are DevOps certifications

- top retail companies in Japan

- top SaaS companies Germany

- competitors of Glaxosmithkline

- Salesforce hq

- Amazon growth rate

- software testing skills

- requirements for CISCO certification

- RN salaries in Texas

- New Jersey area codes

- Kansas City zip codes

- common Latino last names

- women's names in India

- how many members on reddit

- how old is Meryl Streep

Google also recognizes queries in foreign languages. It often (though not always) includes terms' translations to English in the search results. Example: питон линкедин (a query in Russian).

Google may provide "the" answer to a query in a language other than English (say, in Russian.) Examples:

- как далеко до луны

- погода завтра

- кафе неподалеку

- сколько от москвы до санкт-петербурга

- сколько лет генеральному директору гугл

If you have a question that you can comfortably put in one sentence, Google will likely know a satisfying answer. Use simple queries, for instance, to investigate terminology or to find target companies for further research.

For example, we would run queries like top US transportation companies before looking for transportation professionals to identify the major industry players. But to proceed further, you need to use a different way of thinking and searching for the cases when your target cannot be identified by one question.

THE MAIN SEARCH PRINCIPLE – "VISUALIZE SUCCESS"

Google displays web pages that include the terms you enter. To search productively, you should use keywords (and key phrases) that you *expect to find* on the resulting pages. As a past version of Google Help reads,

"Choose words that are likely to appear on the site you're looking for."

Think of what a person could have written about themselves or what others could have written about them. Identify and use such keywords and phrases that your target pages have, and other pages (the ones of no interest to you) do not contain. Then you will get the desired results faster. We call the search principle *"Visualize Success."* Start thinking in this manner, and your results will significantly improve. Once you begin magically encountering superb results (like never before), you will understand how to think about the terms to type into the search box.

The following strings demonstrate searching by that principle: *imagining* the words and phrases you will find in the results and then constructing search strings using that text.

Examples:

- "is a software engineer at Google" – most results are pages talking about someone and naming that person's job title and employer.

- "is a member of women in technology" – each result is a member of WITI (likely, a woman). These two examples represent what specialists call "Natural Language Search."

- vp finance "earned her MBA from" – results will likely contain womens' names for VP Finance with an MBA.

- site:linkedin.com "join to connect" - the phrase "Join to connect" is part of every public English language LinkedIn profile and is rarely found elsewhere on the LinkedIn site. So, searching for the phrase will narrow results down to LinkedIn profiles. You will not see the LinkedIn company, group, articles, jobs, or other types of pages in your results.

- association recruiters "local chapter" California – a page of an association chapter site may contain these words.

- "back to staff directory" – a page that says this will likely lead us to an organization staff directory.

Using words and phrases that you expect to find within results helps you to surface excellent results. Compared to simple search strings with keywords and phrases, strings with additional Boolean syntax elements provide further control over search results. For example, you could be looking only for files of a given type or only for files on a given site. We have included detailed descriptions of Google's Boolean syntax and operators below.

Searching for contact lists is another excellent application of the "Visualize Success" principle. You are more likely to find contact lists by searching for:

- several email domains, including "gmail.com"

- country email extensions like "uk" and "fr"

- phone area codes

- postal codes

- common first or last names

- job titles

- certification abbreviations,

than if you "pose a question" such as <u>email lists of big four consultants</u>.
Example search:
<u>association "registered nurses" "gmail.com" "yahoo.com"</u>
– naming what we expect to find: a page on an association site containing some email addresses. Using two email extensions in the quotation marks as keywords, we hope that we will discover contact lists). Compare results with <u>contact lists for registered nurses association</u>. The last search will take you to sites that sell lists – not what you wanted.

Companies with email domains that are different from site domains are especially easy to look up in contact lists. Some separate email domain examples include the following:

- fb.com

- us.nestle.com (and similar domains for countries)

- us.pwc.com

- us.ibm.com

- us.panasonic.com

- ra.rockwell.com

- yahoo-inc.com

- pge-corp.com

If you search for any of these values, you are likely to find employees' email addresses.

Searches that will find pages with contact emails of these companies' employees include the following:

- <u>"fb.com" senior recruiter</u>

- "ca.ibm.com" "management consultant"
- "pge-corp.com" officers
- "ca.ibm.com" "yahoo.com" "gmail.com" mike lee

Here are other contact-list-finding examples, benefiting from looking for more than one email domain or phone area code (by the way, you can search a postal or phone code, and Google will show you the relevant locations):

- "raytheon.com" "lmco.com" "ngc.com" defense research
- members "nl" "co.uk" "de" +44 +31 +49 textile packaging materials
- members "nl" "co.uk" "de" +44 +31 +49 Europe pharmaceutical
- site:marc.info "redhat.com" "gmail.com"
- "astrazeneca.com" "merck.com" filetype:xlsx
- email filetype:PDF "bankofny.com" "wellsfargo.com"
- "sorted by name" "gmail.com" "us.ibm.com"
- "pfizer.com" "novartis.com" "merck.com" "roche.com" David Pat senior manager
- Michigan "accenture.com" "deloitte.com" "vendor activity"
- 206 425 "accenture.com" "bah.com"
- +91 +66 "Novartis.com" "Pfizer.com"
- +65 +62 +66 "th" "sg" "id" pharma phone list
- "gmail.com" "hotmail.com" "aol.com" mining geosciences alumni
- "exxonmobil.com" "chevron.com" manager government relations David Barbara
- "deloitte.com" "accenture.com" 415 650 408

(We have included so many search strings for you to experience remarkable results for each and to convince you to use the technique.)

Any list is worth exploring if you think it may contain even a few contacts of interest.

When you search for contact information, you might want to limit results only to the recent two or three years since contacts get outdated fast – especially work emails. (Search with no date restriction too.) But if you find "gmail.com" or other free email domain-based addresses, people usually keep them for life. We explain how to limit results to a time interval later in this chapter.

Finding email lists is a favorite sourcing method because an email address points to a person you can identify, along with the name, location, and social profiles, by using available tools. These tools identify lists of people by email addresses alone, without a need to supply any other information. For example, you can find registered LinkedIn profiles by uploading an email list to your account. (You can do it regardless of your subscription.) All you need is email addresses. You will see all profiles of LinkedIn members who have used an included email in your list.

Tools like pipl.com, Contacts+, Clearbit, People Data Labs, and others have databases of professional and social media data. None of the tools provide very good precision, especially for phone numbers, so you need to run some sort of output verification to make sure you do not end up contacting the wrong people. But these tools can work with lists. You can query them in bulk by uploading lists of emails. You can download the output in Excel format. It will typically contain names, employers, job titles, and links to social profiles if those are present in their databases. (Most tools charge the user based on the number of returned results.) Some tools also try to get results dynamically, based on your query, by running and processing output from some search engines and systems. They might be mistaken in vague user input cases more often, but you can be sure the results are at least as up-to-date as Google's index.

To find results for sophisticated queries, you should learn how to use advanced Google operators.

SURFACE WEB

When you click the "search" button, Google (or any other web search engine), the results you get back are *not* based on a real-time snapshot of the web as it is at that moment. Instead, the results come from the search engine's index. A search engine's index is much like the index an author provides at the end of a book.

For many search terms, Google has a list of the hundreds of billions of web pages containing that term. Whenever Googlebot visits and gathers information about a web page, it updates the index for every term in the page. This is also true of other information, like structured data (such as Schema.org objects, Meta tags, and microformats) in the page.

Google has determined that some pages change frequently (like the front page of major news outlet websites), while others change infrequently (like many of <u>these search results</u>). Googlebot optimizes the frequency of revisiting pages accordingly. Google revisits (re-indexes) some pages many times a day and others only a few times per year. Therefore, search results may show an older version of a web page that is not up to date. A page may be gone, but a *cached* copy (a page copy stored in the index) is still available. Often both the cached and live versions of pages are of interest. Access to cached pages gives us an advantage in research.

What can you find by searching on Google.com? Can you find any page on the web? We sometimes ask our attendees this question, and it takes a minute before someone responds – often incorrectly.

A factor to consider when determining what Google can find is that search engine crawlers like Googlebot cannot visit every page on the web. Website owners can submit a list of pages on their websites for Googlebot to visit. They can prevent crawlers from accessing their pages. To visit a web page, Googlebot must have permission to visit the page. Googlebot may also discover pages by following hyperlinks (just like a regular user would).

Unlike a person browsing the web, Googlebot follows some well-defined rules about what pages it can and cannot copy to its index. Links on the web contain invisible metadata that tells Googlebot (and other crawling software) what links to follow or not to follow and what information it can index from which pages.

Most websites contain a special file called *robots.txt* that tells web robots (like Googlebot) which pages it can and cannot visit or index on that specific site. Finally, Googlebot (obviously) does not log into websites. It does not have user accounts on Facebook, LinkedIn, or the American Institute of Certified Public Accountants' website. So, if a user must log in to view a specific page on any website, Googlebot will not be able to view that page or index it.

All of the web pages that Google can visit and index are collectively referred to as the *Surface Web*. All of the pages that Google cannot visit, or index, are collectively referred to as the *Deep Web*.

The deep web is much more extensive than the surface web. Take a look at some graphical representation of surface vs. deep web: surface deep web. The two terms are usually represented as parts of an iceberg (Figure 1.1):

FIGURE 1.1   Illustrating data available on the deep web vs. surface web.

While many times smaller than the deep web, the surface web is still enormous. Google does not disclose the total number of pages available via Google search, but it supposedly has 100 million gigabytes of data and 30 trillion pages in its index. That is certainly enough information for us to pay attention!

BASIC BOOLEAN SEARCH SYNTAX

Boolean search means using the Boolean logic to combine search terms with AND, OR, and NOT in the ways a search engine recognizes.

Use the Boolean logic AND and NOT to narrow results and OR to expand them. After running an initial search:

- Add keywords to get fewer, more targeted results – AND

- Remove keywords to exclude unwanted results – NOT

- Vary search terms to get more results – OR

- Put the quotation marks around phrases (sequences of keywords) to narrow the search to pages with the exact phrase, get fewer results

As some examples, you can search for professionals who are

- Engineers but NOT Managers AND NOT Recruiters

- Employees of Google OR Facebook OR Microsoft

- Have a Bachelor's OR a Master's Degree in Physics OR Computer Science

- Living in the Bay Area but NOT in San Francisco

- Have a job title containing the phrase "Call Center Manager"

While Google supports Boolean logic, you cannot search using the capitalized "NOT" to exclude terms. Make sure you follow the correct Google syntax for the Boolean search. For exclusions, use the minus (-) right in front of a word or a phrase.

- OR statement - OR needs to be capitalized: *scientist OR researcher*

  - Alternative OR syntax - *scientist | researcher*

- AND statement - no operator (don't use AND in search) - everything is combined by default: *profile LinkedIn Chicago mathematics*

- NOT statement - the minus right in front of a word: *-recruiter -president -director*

- Put phrases in the quotation marks to find them exactly: *"big data"*

Unless you put a word inside quotation marks, Google automatically also includes pages containing variations of the word, synonyms, and related terms in the results. Because of that, you do not need to use OR statements for synonyms (e.g., *developer OR "software engineer" OR coder*). In fact, if you use an OR statement, Google will no longer look for synonyms – so, controversially, with ORs, you may get *fewer* results than without them. It is unwanted behavior. We recommend against using ORs on Google.

On Google, the order of Boolean operators is predetermined. The operator OR always has the highest priority. You could include OR statements in parentheses – e.g., *(apples OR oranges) bananas* – but the search will be equivalent to *apples OR oranges bananas*. Google will look for pages that contain either "apples" or "oranges" and "bananas."

ADVANCED SEARCH OPERATORS

To start reviewing how advanced search operators look, use Google's Advanced Search Dialog (Figure 1.2):

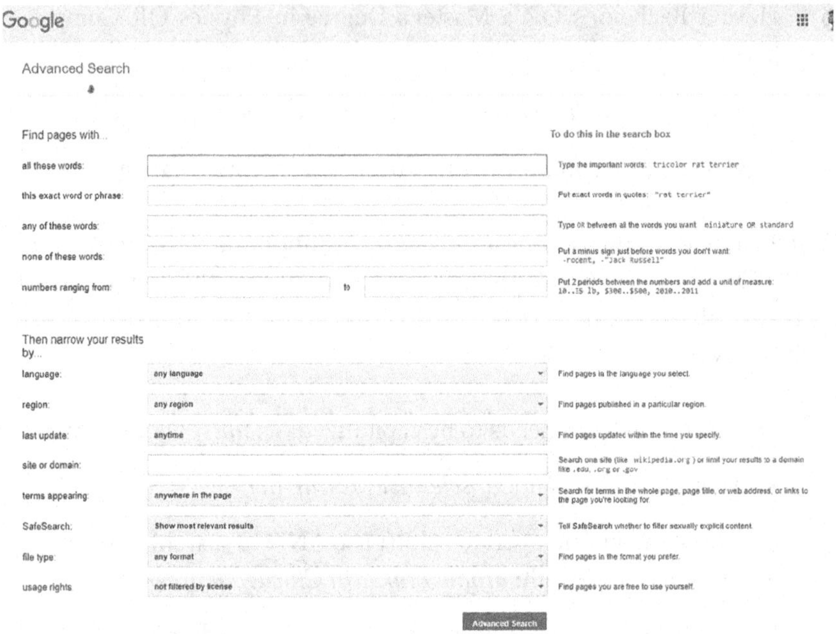

FIGURE 1.2   Google's Advanced Search dialog.

You can enter values in the fields

- all these words: (AND logic)
- this exact word or phrase:
- any of these words: (OR logic)
- none of these words: (NOT logic)
- numbers ranging from:
- this exact word or phrase:
- any of these words:
- type OR between all the words you want:

to set the Boolean logic for your query after you press ENTER.

These values entered in the advanced dialog generate Google advanced search operators (*site:, inurl:, intitle:, intext:,* and *filetype:*):

- site or domain:

- terms appearing:
- filetype:

Once you press ENTER, you will see the correctly constructed search string with operators.

Note, however, that it is not practical to continue using the advanced dialog because you can only generate a limited range of searches with it.

Google has other search operators, which do not appear on the advanced dialog, but these are the most useful ones for research. (We have included a complete operator list in Appendix X). Note that Google's help no longer documents most of the advanced operators. So, the majority of people who Google are not aware of them.

Here is how advanced operators affect your search results. You can combine them with each other and with keywords.

site: (also called X-Raying) – look for results from a given domain

- site:edu – search websites that end in edu (i.e., educational organizations)
- site:nih.gov – search the National Institute of Health website
- site:ie.linkedin.com – search LinkedIn Ireland
- site:stackoverflow.com/users – search for Stack Overflow user profiles

intitle: Search for keywords or phrases in the title of a web page (the blue text in Google search results):

- intitle:"about us" – the phrase "about us" must appear in the page title
- intitle:"member directory" – the phrase "member directory" must appear in the page title

inurl: Search for keywords or phrases in page URLs:

- inurl:careers – the word "careers" must appear in the URL
- inurl:directory faculty earth environmental sciences – adding keywords to inurl: search

intext: Search for keywords or phrases in the text of a web page (not in the title or URL).

- intext:gmail.com

filetype: (can also be written as ext:) – search for a specified file type, such as PDF, DOC, TXT, or XLSX:

- filetype:pdf resume engineer – PDF files
- filetype:xlsx contact list – Excel files

Operator and keywords combinations:

- intitle:"member directory" clinical research professional association – This search will find web results with the exact phrase "member directory" appearing in the web page titles. Further, results must contain the keywords *clinical, research, professional,* and *association* in either the title, URL, or body of the page.

Google may also return results that contain alternate spellings, variations, synonyms, or related words for any of those four keywords. The expectation would be that results would include member directory listings of professional associations serving clinical researchers.

- intitle:chevron site:businessinsider.com 2020 – A search for pages with *Chevron* in the title, from the domain businessinsider.com, containing the number *2020*. Here, you would expect to find articles about the corporation published by Business Insider in the year 2020.

- inurl:orgchart internal audit – A search for pages with "orgchart" in the URL that contain both terms *internal* and *audit* (or alternate spellings, variations, synonyms, or related words) somewhere on the page. This search represents an attempt to find organizational charts posted online that include listings for internal audit staff.

- Other examples:site:usgbc.org/people intitle:leed ap "green building council"

- site:crunchbase.com/person intitle:UX intitle:designer san francisco

- site:contactout.com intitle:"business analyst"

- site:doximity.com/pub intitle:urology intitle:"new york"

- inurl:authors site:fossies.org

- site:xing.com/profile intitle:"regulatory affairs"

- site:specialtyfood.com/organization packaging

- site:ohio.gov insurance agent verify license

- site:apua-asea.org filetype:pdf liste

- site:com/about minority-owned

- site:gov "do not distribute" 2020

- site:uk intitle:"delegate list"

- site:constructionequipment.com/company e-mail

- site:npidb.org/doctors/pharmacy/pharmacist "medical center" OR "university hospital"

- site:chrome.google.com email extractor

- site:reuters.com/finance/stocks/officer-Profile "human resources"

- site:youracclaim.com/users Cisco

- site:facebook.com filetype:smith

- filetype:pdf "advanced search" Google Boolean tips library site:edu

- site:napw.com/users certified financial planner

- site:amazonaws.com intitle:"management team"

- director marketing filetype:xlsx name company title email

- site:research.google.com/pubs author machine learning neural vision

- site:dou.ua/users java back-end

- site:prweb.com healthcare appointed CEO

- "voir ce profil dans une autre langue" "Région de Paris , France" site:fr.linkedin.com

- email format deloitte.com

- site:apha.org roster

- site:angel.co/p "ios developer" bay area

- site:espeakers.com/marketplace/speaker/profile women in business

SEARCH OPERATOR ASTERISK * – "FILL IN THE BLANKS"

We would like to introduce one more Google operator (or, rather, search modifier), the asterisk *. In some search systems, the asterisk replaces part of a word. It works differently on Google – it stands for one word or a few words.

An excellent application of the asterisk is to look for poems or song lyrics for which you may have forgotten some words.

Example: "I * lonely * cloud."

You can use more asterisks if you forgot two or more words: "we all * * * submarine."

In research, you can use the asterisk to collect pages with similar information that only differ in some words. Examples:

- "email me at * * com"

- "joined * as chief * officer"

- "senior * engineer at * inc OR llc"

- "have * years of experience"

These searches apply the idea of Natural Language Search, with the convenience of skipping some words within a sentence.

INCLUDE OMITTED RESULTS

An important setting that you need to be aware of is *"include the omitted results."* By default, Google shows only representative results. You can include the omitted results either by clicking a link on the last page of results that says "repeat the search with the omitted results included" or by adding *&filter=0* to the search URL. (You will still get up to 300–500 results, not "all.")

When you are looking for a quick answer, including omitted results is not necessary. But if you are looking to see as many results as possible, do use the setting.

GOOGLE IMAGE SEARCH

Once you enter your query into the box and press ENTER, you can switch to specialized searches: Images, Maps, Videos, News, Shopping, Books, Flights, and Finance.

In addition:

If your search implies jobs, colleges, or events, Google will take you to those specialized searches (which do not have dedicated web pages or links on the search page).

Further,

- if it is about services, Google will show a map

- if you are researching a product, you will see shopping sites

- if you are looking to travel, you will get a flight search page

- if you are looking for a song, you will see videos

The image search is quite useful in research. Note that it does not include any promotions, and it is not a place where most sites compete for visibility. Therefore, with image search, you may unearth relevant results that the "All" search will not show until the second or third page, if at all.

Unfortunately, recently, the image search has lost some parameters: if you search on Google and switch to Images, some previously available options will be missing. Use the <u>Advanced Image Search</u> to be able to enter every possible parameter. The image search also has its own operator *imasize:*.

Image search supports narrowing results by the following:

- Image size

- Aspect ratio

- Color

- Type (such as face or photo)

- Region

- From site

- Image format

- Usage rights

It is quite practical to be searching for *faces* if you are looking to find people and information about them.

You can search for different image aspect ratios in subsequent searches to see new results.

Here is an example of using image search from our practice. We were looking to find Flight Test Specialists, Safety Engineers, and Propulsion Engineers with experience at a company that makes passenger aircraft. But we were not familiar with the players. Searching by a company name showed us either passenger planes or military planes, small planes, engines, or helicopters. If we needed to narrow down the search, we would look for the "photo" types. Those companies that displayed helicopters, for example, were not our target. Comparing the screenshots on the left and right (Figure 1.3), it took seconds to assess each company:

FIGURE 1.3 Comparing image search results to determine a company's products.

Another use case was searching for employees for a company that develops and tests drugs. Images returned by searching for a company name showed either testing equipment or listed software testing features and computer screens. We needed companies that make both equipment and software and could pick them from the image search results by choosing photos with testing equipment.

NOTES FOR PRACTICAL SEARCHING

Important Note on Keyword Variations. For keywords (without the quotation marks), Google looks for synonyms and variations. However,

- For a word used with the minus -, only this exact word is excluded (but not its synonyms or variations)

- For a word following an advanced operator (*intitle:*, *inurl:*, etc.), the exact word is used (no variations)

There is no way to exclude all synonyms of a word using the negation operator minus (-). If you need to be thorough, you have to think about possible variations or spot them from the results page and adjust the search by excluding additional terms. Add exclusions until your results have mostly "correct" pages.

Word Order. The word order matters in searching. Google gives higher ranks to pages with the same word order as in the search string. (So, when you search, try putting your keywords in the order you expect to find them – e.g., type *software engineer*, not *engineer software*.) Pages containing the entered keywords in close proximity and the same word order (but not necessarily one after another) will also be ranked higher. If you type a long phrase into the search bar, the top results will typically include the phrase, even if you do not use the quotation marks.

Changing the word order in a search string can change the order of results and affect the number of returned results.

The order of advanced operators within a search string is less important than the keyword order. (But it may affect the order of results as well.)

Auto-Corrections. Pay careful attention when Google suggests spelling corrections and other changes to your search. While helpful in avoiding common typos, this feature often misfires on uncommon (but correct) words and technical terminology.

Note that Google usually wants to auto-correct the spelling *e-mail* to *email.* But searching for "e-mail" sometimes provides more results (which also include pages with the word "email").

NUMBER OF RESULTS

In theory, Google never shows more than 1,000 results. Try to find results beyond 1,000, and you will see a message from Google about the maximum number of results (Figure 1.4):

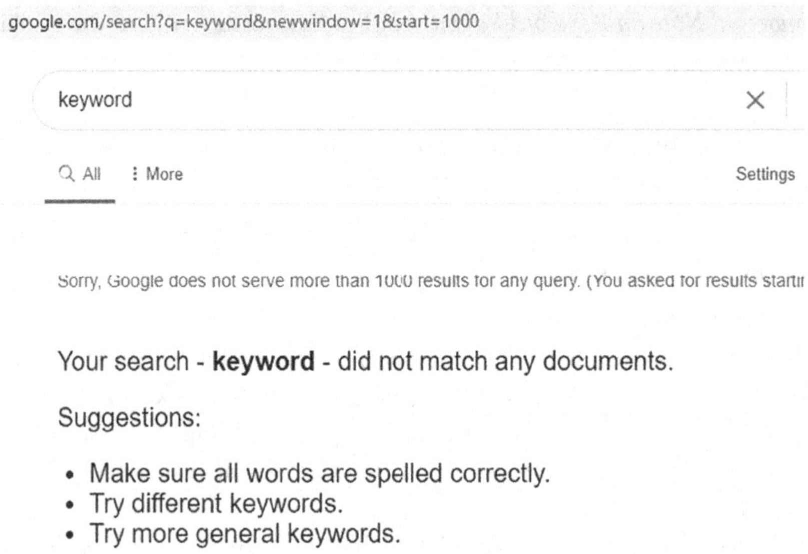

google.com/search?q=keyword&newwindow=1&start=1000

keyword   ×

Q All   ⋮ More   Settings

Sorry, Google does not serve more than 1000 results for any query. (You asked for results startir

Your search - **keyword** - did not match any documents.

Suggestions:

- Make sure all words are spelled correctly.
- Try different keywords.
- Try more general keywords.

FIGURE 1.4   Demonstrating Google's cap of 1,000 search results.

Google used to show up to 1,000 results, but with the index volume getting bigger and various algorithm modifications, it no longer does. However, in most cases, Google provides fewer results – it "maxes out" at 300–500. We do not know of a Google search that currently delivers over 500 results.

Google's displayed number of results is often off by orders of magnitude. (It's just not a high priority for Google to display a number that closely reflects the total number of results in Google's index). It is a matter of constant user confusion. Some users compare numbers of results for different searches, but it is impossible to do so since the numbers are removed from reality.

Instead of the unreliable number, we wish Google would display a qualitative characteristic like "many results."

The number of results is not to be trusted on any global search engine.

CROSSED-OUT WORDS IN RESULTS (SOFT "AND")

In some cases, where Google can't easily find pages containing all the keywords, it provides suggested results in which one or more keywords are missing. It does this because it wants to give at least some results for any query.

Google shows crossed-out keywords below the search results in cases like these. If you want a word to appear in the results for sure, you can put it in quotation marks and search again.

SEARCHING VERBATIM

You can select the *Verbatim* option under the search bar after you choose the *Tools* button. It instructs Google to provide no interpretation of any kind of a search string. Using Verbatim will result in Google looking for the keywords precisely as they are, not their variations or synonyms. However, sometimes, Google may still decide to show some results with crossed-out keywords.

SEARCHING BY DATE RANGE

You can choose a date range for the search results. Google provides a calendar dialog for that when you first search and then click the "Tools" button.

In 2019, Google introduced the operators *before:* and *after:* to search within a date range.

If you choose a date range, Google will display the last updated dates in the snippets. To see the latest results, you can restrict your search to the last minute or even a second.

While restricting the dates to more recent may seem like a good search strategy, you should keep in mind that a page "date" is often technically hard to identify. It appears that Google has many pages in its index for which it "doesn't know" the date. (Many LinkedIn profiles have this quality.) This happens due to pages not having enough information about the updated dates.

Some sites attempt to "hack" the date range by modifying the date without changing the content. When you set a date range restriction in Google search, you will stop seeing matching results with no specified date.

Our practical advice is to only search by a date range when looking for information that strongly correlates with a date or period of time, such as an event. For most searches, you do not need to utilize the date range; Google automatically ranks higher newer results.

CHAPTER 2

What Is a Google Custom Search Engine?

G OOGLE CSEs SEARCH THE surface web, just like Google does. They, too, search within <u>Google's Index</u> (data collected by Googlebot) in particular ways, defined for each CSE by its creator.

The purpose of CSEs is to provide end-users with custom ways to search within Google's Index.

Google's CSE help is primarily oriented towards website owners wanting to offer their site search for visitors. Many sites use this technology. Here is a <u>list of over 500K domains that use CSEs</u>. However, CSE applications go way beyond your own site search, allowing you to search sites built by others from which you want to extract information. That is our focus.

A BIT OF HISTORY

Google introduced CSEs back in 2006. A blog post from one month after the launch, highlighted the rapid adoption of CSEs:

> Less than four weeks after Google launched its Custom Search Engine, a tool to create customized search engines, the Net is being flooded by search boxes stamped with the phrase "Google Custom Search."

The post also had a screenshot of the original CSE interface (Figure 2.1):

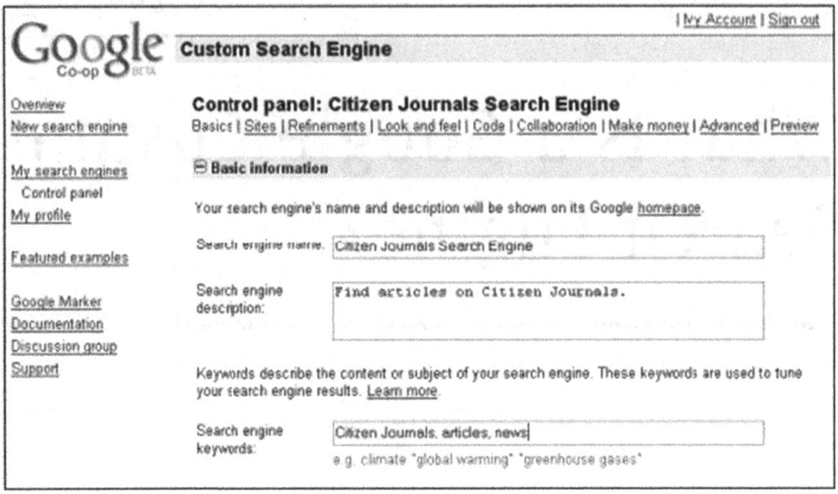

FIGURE 2.1   The original CSE control panel.

Apparently, most of today's CSE settings were already present in the first implementation. The CSE Control Panel UI (unfortunately) hasn't significantly changed either. In recent years, Google has added powerful semantic search features to CSEs via the language of Schema.org and Knowledge Graph objects. (*Semantic* means recognizing the user's intention.)

We will tell you all about that.

CSEs VS. GOOGLE – ADVANTAGES AND CHALLENGES

Why would you search on a CSE vs. Google.com? The notable CSE advantages are as follows:

1. CSEs can hide complex or repeated search syntax from the end-user.

On Google, you can search for site:linkedin.com/in (plus keywords) and get only pages from linkedin.com/in (which are LinkedIn public profiles). You would need to enter the operator every time you search.

You can instead set a CSE to append a search string to each end-user's search automatically. As an example, you can hide the *site:linkedin.com/in* operator from the end-user and invisibly add the string to every search.

Or you can invisibly set a CSE to only return PDF files by automatically adding the string *filetype:PDF*.

These are useful, time-saving features, but CSEs offer much more.

2. CSEs prevent Google from displaying CAPTCHAs.

If you use advanced Google search operators, you are well familiar with annoying CAPTCHAs. The CAPTCHAs intended use is to prevent unauthorized massive Google use via scripts, but they affect your research productivity (Figure 2.2).

FIGURE 2.2   A sample Google search CAPTCHA.

With CSEs, you will never need to solve those puzzles that slow you down.

3. CSEs offer advanced settings unavailable in Google.com.

For example, you can include or exclude pages and rate pages higher if they have particular keywords or are on a specific site. You can even set the weights with which particular sites are included, prioritizing some sites over others. You can also not only restrict but boost results by country.

4. CSEs offer unique advanced search filters to more technically advanced end-users.

CSEs do "understand" advanced Google Boolean search syntax, including operators (such as *site:*). They also "understand" little-known CSE-only search operators. The operators allow end-users to run filtered searches against a page's "structure," which is hidden in the page source. With the operators, you can run filtered searches across some websites or even the whole web. For example, you can search for *LinkedIn profiles of Tesla employees* or *GitHub profiles of San Francisco residents*.

Custom Search Engines' drawbacks include the following.

For creators and advanced end-users:

1. The CSE Control Panel is confusing. They have not updated the UI forever!

2. The documentation is sparse. It is especially sketchy on the topic of writing advanced CSE search operators.

3. Figuring out how to write CSE operators is tricky and somewhat technical (it requires you to "read" JSON-formatted files).

For end-users:

4. CSEs restrict results to 100 in one search. (With Google, you can get up to 300–500 results per search.)

While there is no easy way to overcome "4," we hope that this book will help with issues "1," "2," and "3"!

CHAPTER 3

Creating Your First CSE

G OOGLE.COM HAS ADVANCED SEARCH operators that help to control results. The operator *site:<site>* tells Google to narrow down results to only include those from the *<site>*. *<site>* can be *com, edu, de, org, uk. linkedin.com, support.microsoft.com,* or *purdue.edu/advance-purdue/ conferences.*

(Replace the terms in brackets <> to run searches.)

For example, this search

site:linkedin.com/in javascript engineer "san francisco"

will find pages with URLs starting with *linkedin.com/in* – which are public LinkedIn profiles – containing the rest of the keywords (Figure 3.1):

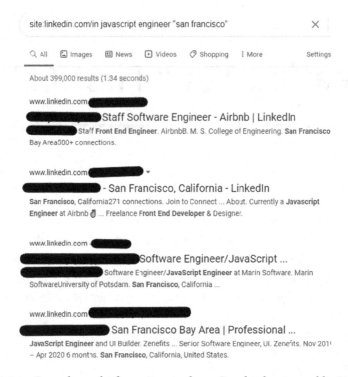

FIGURE 3.1   Example results for a *site:* search on Google, showing public LinkedIn profiles.

Custom Search Engines (CSEs) is a tool that queries Google in specific ways. A CSE can, for example, include some search terms but hide them from the end-user. A CSE, the end-user can, for instance, skip typing in the *intitle:* or *filetype:* operators.

We will create a simple CSE that will execute but hide the *site:* operator step-by-step.

To define a CSE, you must first log in to your Gmail account. If you are not logged in, you will see a "Sign in with your Google Account" dialog. If you do not have a Gmail account, you will need to create one. Press "Create account" on Gmail.com and follow the prompts.

You can find the necessary instructions for CSE creation described in Google's documentation, but we are going into much more detail in this book. (Note that Google's documentation addresses you as a website creator. It is not our use case; we are going to search sites that others have created.)

The URL for you to start with, the *Dashboard*, is https://programmab-lesearchengine.google.com/cse/all.

If you have never created a CSE, your list of CSEs will look empty (Figure 3.2):

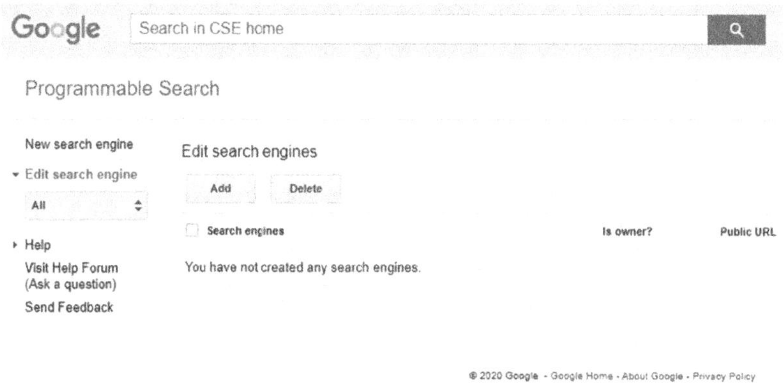

FIGURE 3.2   The CSE creation and editing interface before creating your first CSE.

(The search box on the top of the page looks for keywords in Google's CSE Help.)

Press the "Add" button to start defining your first CSE. You will see the creation dialog (Figure 3.3):

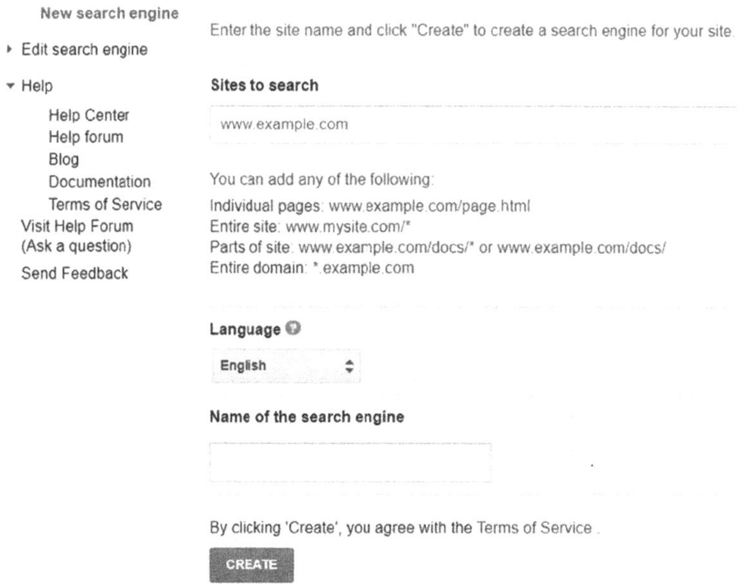

FIGURE 3.3   The CSE creation interface, steps to create your first CSE.

You must add at least one site to search (for example, *linkedin.com/in*), select the UI language if different than English, name the CSE, and click on the "create" button.

CHAPTER 4

Editing and Testing Your CSEs

I N THE CONTROL PANEL, you can navigate to editing various CSE settings by choosing a section on the left and then selecting a tab at the top
of the page (Figure 4.1):

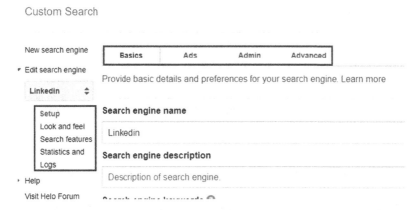

FIGURE 4.1   The CSE Control Panel showing the top of the Setup screen's Basics
tab, highlighting navigation to other tabs and screens.

Note that most parameters take effect as soon as you have changed them (and there's no "undo!"). However, some settings – inconsistently – require you to click a "Save" button. Watch for those to make sure that your new entries take effect.

Settings with the most useful functionality, which we will discuss in detail, are in these Control Panel sections:

- Basics -> Setup – all fields, including the Sites to search

- Search features -> Advanced -> Websearch Setting – "Query Addition"

- Search features -> Refinements -> Synonyms

Please review all the CSE settings by clicking on the left tab and selecting each of the top tabs.

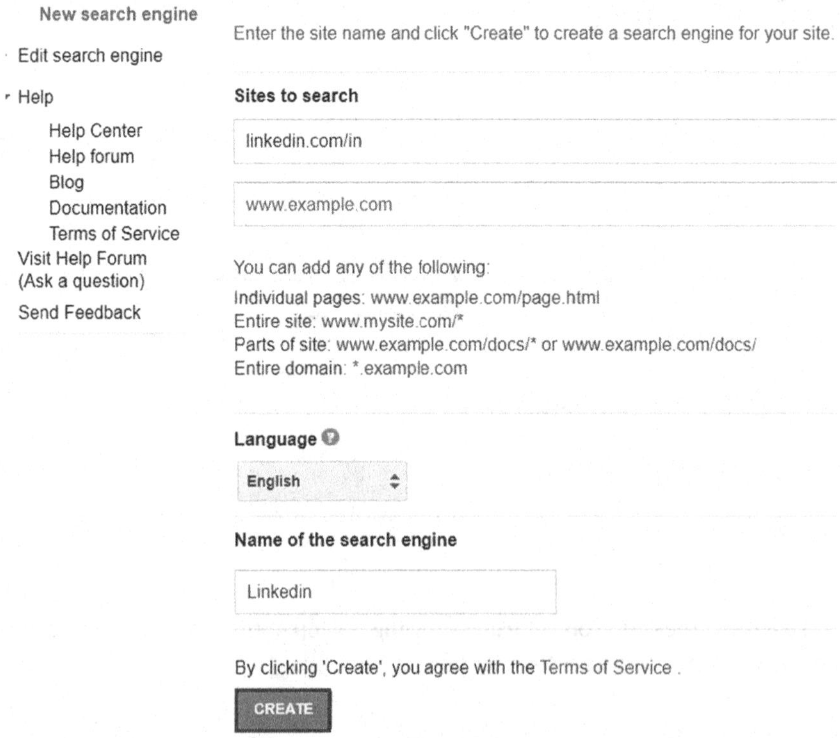

New search engine

Edit search engine

˒ Help

Help Center
Help forum
Blog
Documentation
Terms of Service
Visit Help Forum
(Ask a question)
Send Feedback

Enter the site name and click "Create" to create a search engine for your site.

Sites to search

linkedin.com/in

www.example.com

You can add any of the following:

Individual pages: www.example.com/page.html
Entire site: www.mysite.com/*
Parts of site: www.example.com/docs/* or www.example.com/docs/
Entire domain: *.example.com

Language ⍰

English ⇅

Name of the search engine

Linkedin

By clicking 'Create', you agree with the Terms of Service .

CREATE

FIGURE 4.2   The CSE Control Panel, naming and creating your first CSE.

After you have pressed "Create," you will arrive at this screen:

FIGURE 4.3 The CSE interface, post-CSE creation, highlighting navigation to the Control Panel.

The CSE is ready to run. You can access it by pressing the "Public URL" button. Try running several keyword searches and examine the results.

Now, let us see how we can affect its search results by manipulating various Control Panel parameters.

Press the "Control Panel" button and proceed to define the CSE parameters. The full Control Panel screenshot won't fit on one screen, so we will go through the available CSE parameters in two steps.

Here is what you will see in the Control Panel on Setup (on the left)/ Basic (on the top):

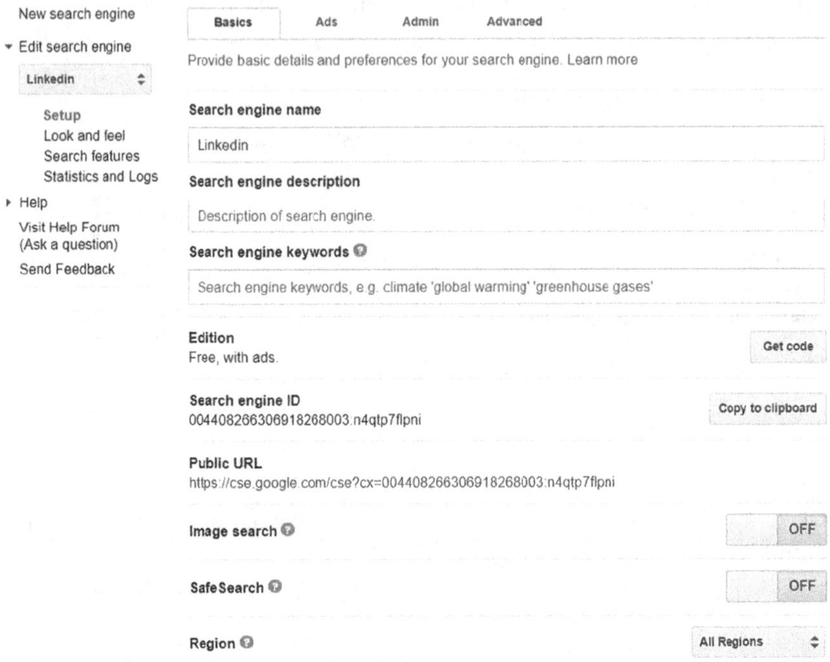

FIGURE 4.4   Top half of the CSE Control Panel Setup page, Basics tab, for a newly created CSE.

Search engine name and description play no role in how it searches and are invisible to the end-user. They are for the creator's use only.

Search engine keywords influence ranking. It is a "soft" setting. This field is optional. If you fill it out, the CSE will include pages without the keywords but will give those with the keywords a higher rank. Giving keywords a priority is not possible on Google.

The *Edition* setting may show any of *Non-profit, ads optional*, *Free, with ads*, or *Standard* and possibly other variations. The Edition reflects the status of ads in your Custom Search Engine. Unless the *Edition* is *Non-profit, ads optional*, your Google CSE will display advertisements to the end-user, just like regular Google search does. Under the Setup -> Ads page of the Control Panel, there are options to enter your non-profit ID or connect your AdSense account. If you provide proof of your non-profit status, you can tell Google not to display ads on your CSE. With a connected AdSense account, you can earn money when end-users click on advertisements Google shows on your CSE.

Google generates a unique *Search engine ID* and *Public URL* for you and other people to run searches. You can copy the CSE's Public URL and share it with others.

You can optionally include *Image search*. For the user, the Image search will appear under a separate tab in the CSE.

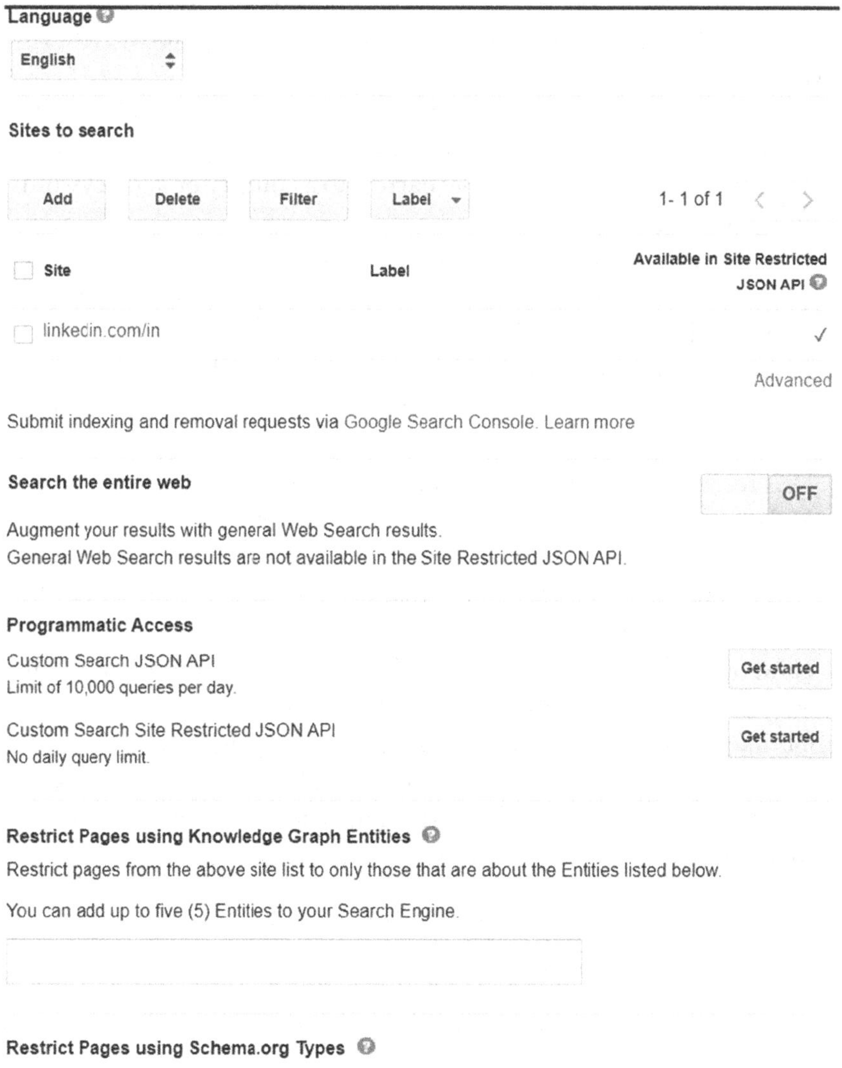

Language ☺

English ⬍

Sites to search

| Add | Delete | Filter | Label ▾ | | 1- 1 of 1 ⟨ ⟩ |

☐ Site	Label	Available in Site Restricted JSON API ☺
☐ linkecin.com/in		✓
		Advanced

Submit indexing and removal requests via Google Search Console. Learn more

Search the entire web   OFF

Augment your results with general Web Search results.
General Web Search results are not available in the Site Restricted JSON API.

Programmatic Access

Custom Search JSON API   Get started
Limit of 10,000 queries per day.

Custom Search Site Restricted JSON API   Get started
No daily query limit.

Restrict Pages using Knowledge Graph Entities ☺

Restrict pages from the above site list to only those that are about the Entities listed below.

You can add up to five (5) Entities to your Search Engine.

Restrict Pages using Schema.org Types ☺

FIGURE 4.5   Bottom half of the CSE Control Panel Setup page, Basics tab, for a newly created CSE.

You can narrow the search down to a country *(Region)*. You can do so in the Advanced Google Search dialog, but in a CSE, you can have the parameter set for all searches without the need to reenter.

Scroll down and look at the second set of parameters:

On Screen Two in the *Sites to search* section, you can specify the site(s) to include in the results and sites to exclude.

You can also select the *Search the entire web option*, and the CSE will include not only results from the listed sites but also from non-listed sites that rank high for the search. Google itself doesn't have a "soft site inclusion" option.

Further, if you select *Search the entire web*, you can remove all the sites to include and search the web like Google. Or, you can use other parameters to define the CSE (for example, narrow to a country or add keywords to affect ranking).

We will set the definition for *Knowledge Graph Entities* and *Schema.org Types* aside for now since they require lengthy explanations. We will cover them in full detail later.

In our example, you started defining a CSE by including a site.

You can enter several sites to include and exclude in the Control Panel. You can also set "Search the entire web" on or off.

Let's recreate the CSE that will search public LinkedIn profiles.

Make sure you have a Gmail account and log in. Start with the Dashboard (https://programmablesearchengine.google.com/cse/all), press the "Add" button, and create a CSE that has the same parameters as the previous screenshot.

Now, test the CSE. Click the Public URL on in the Control Panel and enter a few words – for example, a job title, company name, or location name like "San Francisco Bay Area." Your results will be only public LinkedIn profiles. Here is what a search looks like on our CSE (click the link to reproduce the search): javascript engineer "san francisco". You may need to scroll past some advertisements to see the search results.

FIGURE 4.6   Search performed from the Public URL of the CSE shown in Figures 4.2–4.5.

(The number of search results displayed in the screenshot is misleading, as it always is the case with Google and CSEs. With any CSE, it is capped at 100 results (Figure 4.6)).

RESULTS FOR DIFFERENT END-USERS

Do not be surprised if your results look different from the ones in the screenshot. It can happen because of the following:

1. The search algorithm changes all the time

2. Results depend on your Google settings

3. Results depend on your location identified by the IP address

You can experiment by temporarily changing your IP address using a tool like TunnelBear to pretend that you are in another country. Google and CSE search results will differ for different IP addresses.

Take a look at these two screenshots. On the left, we are searching from home in the US. On the right, we are searching from a Germany-based IP address. In this case, results contain references to Germany (Figure 4.7):

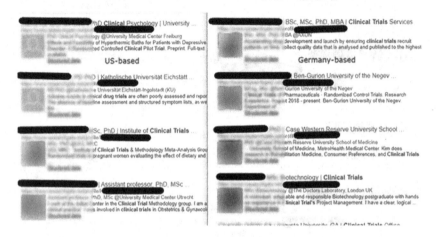

FIGURE 4.7   Comparison of search results using the same CSE with different geo-located IP addresses: left is US, right is Germany.

Go back to the Control Panel and set the CSE to "Search the entire web." Try the same searches. This time, you are likely to see not only profiles but also, for example, job posts. So, this does not seem to be the right setting for a profile search (assuming that was what we wanted). Let's turn the "Search the entire web" off and search only on *site:linkedin.com/in.*

With this first CSE example, you can observe two practical uses of CSEs:

1. End-user does not need to retype the *site:* operator – it is automatically (and invisibly) included.

2. Someone who is not comfortable typing advanced search operators will not need to write them out. The user can simply search for keywords and see the desired results.

Also, notice the narrowing to a country and "soft" inclusions for keywords and sites. You have just started out and already have a wealth of new tools!

You should get used to the Control Panel UI and know what to expect when setting certain parameters and running test searches. It is not hard since the set of available parameters is not that extensive. Play with various values and test the outcomes.

Once you feel comfortable with the UI, you will be ready to easily generate CSEs for instant use for yourself and others (who will surely like you).

Learning where to enter your parameters and in which format is like acquiring ten practical phrases in a local language that would take you to most places in a foreign-speaking country.

CSEs FOR END-USERS

Each CSE has a Public URL that the creator can obtain in the Control Panel and share it with others. The Public URL has the format https://cse. google.com/cse?cx=<CSE ID>, where <CSE ID> is a unique string of symbols identifying the CSE.

A CSE search using a Public URL looks like this (Figure 4.8):

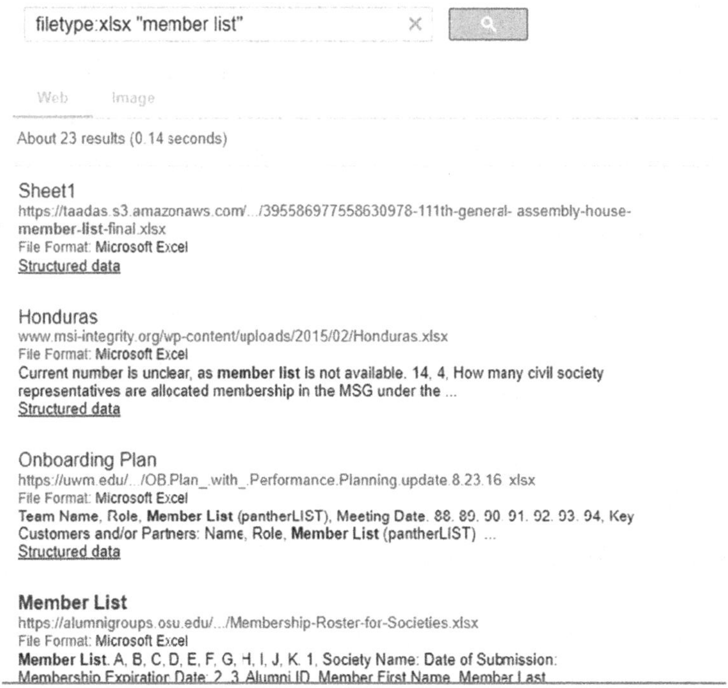

FIGURE 4.8   Example search using a CSE Public URL.

Alternatively to sharing the public URL, creators can include a CSE on any web pages, which they can edit by inserting the provided javascript

FIGURE 4.9 The Get Code screen within a CSE Control Panel, featuring javascript code to paste into a web page to embed a CSE there – an alternative to sharing the Public URL.

code into the page's HTML. Here is what the "Get code" Control Panel page looks like. The only varying part of the code is the CSE ID:

There is no need to write or even understand javascript for the CSE inclusion on a page. You can paste the code into your page editor, save it, and the page will have a CSE box available when it loads. (Note that due to javascript limitations, you can include only one CSE per page.)

A stand-alone, as simple as possible HTML page with a built-in CSE would look like this:

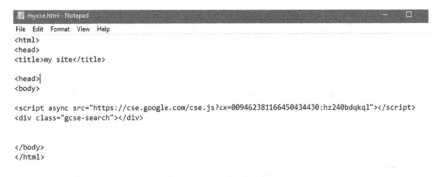

FIGURE 4.10 Sample HTML for creating a stand-alone web page to use the embedded CSE code from Figure 4.9.

When you search using the box on your page, it will look like this:

The CSE end-user can search with keywords and advanced Google search operators, just like on Google (Figure 4.11).

FIGURE 4.11 Example of loading the HTML file from Figure 4.10 in a browser and performing a search.

2

Configure a CSE

CHAPTER **5**

What Are You Looking For?

YOU HAVE BUILT A CSE by entering a site (*linkedin.com/in*) to search. You can now proceed to add a list of sites to include and exclude. (For best performance, Google advises that you not have more than ten sites on your "included" list. If you do, the CSE may start missing relevant results.)

In the next exercise, you will be defining a CSE to search on several popular document storage sites (Figure 5.1).

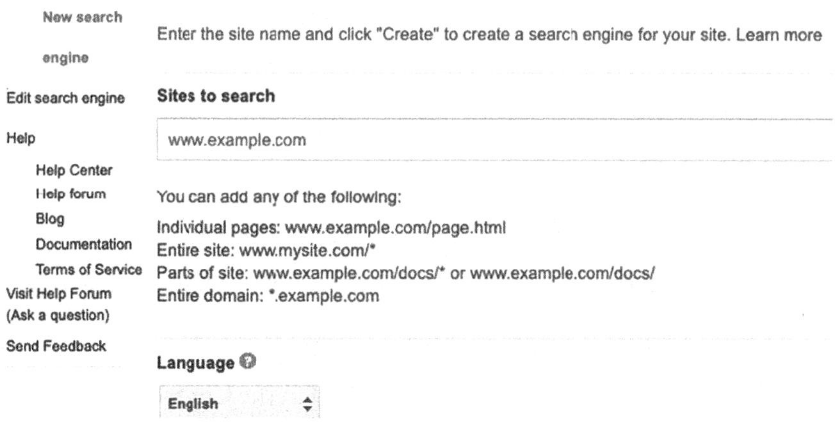

New search engine

Edit search engine

Help
Help Center
Help forum
Blog
Documentation
Terms of Service
Visit Help Forum
(Ask a question)

Send Feedback

Enter the site name and click "Create" to create a search engine for your site. Learn more

Sites to search

www.example.com

You can add any of the following:

Individual pages: www.example.com/page.html
Entire site: www.mysite.com/*
Parts of site: www.example.com/docs/* or www.example.com/docs/
Entire domain: *.example.com

Language ⊘

English

FIGURE 5.1   Creating and configuring a new CSE, step 1.

Start by creating a new Custom Search Engine from the <u>Dashboard</u>. (Do not forget to log in.)

Under the *Setup/Basics* section of the Control Panel, scroll down to see the "Sites to search" section. Here you can add and remove sites from the CSE.

Enter the list of sites below so that your screen matches this image (Figure 5.2):

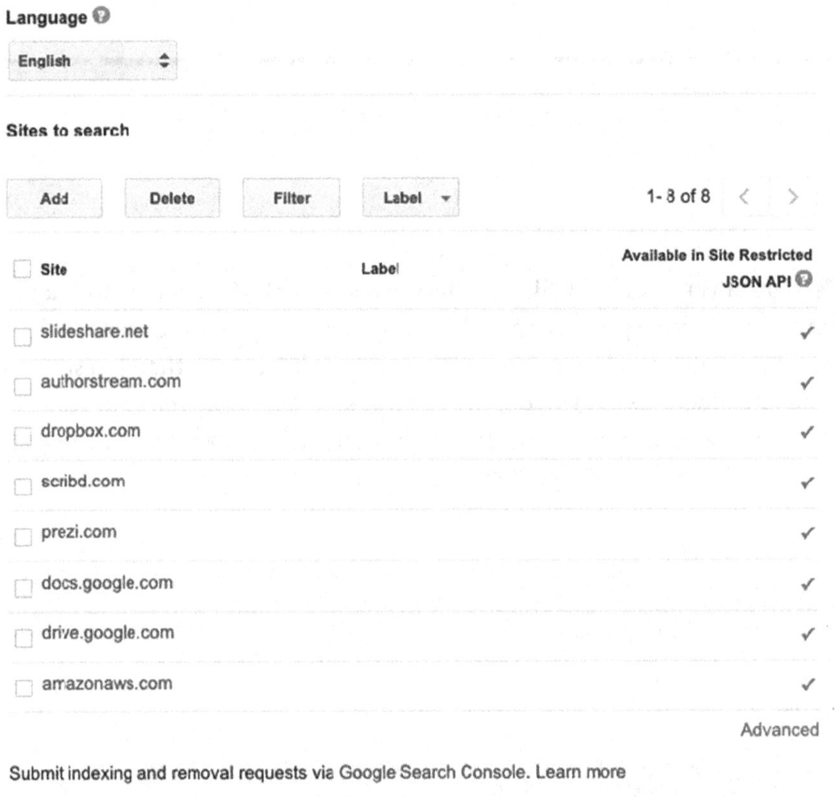

Language ❓

| English | ⬍ |

Sites to search

| Add | Delete | Filter | Label ▾ | | 1- 8 of 8 | ‹ › |

☐ Site	Label	Available in Site Restricted JSON API ❓
☐ slideshare.net		✓
☐ authorstream.com		✓
☐ dropbox.com		✓
☐ scribd.com		✓
☐ prezi.com		✓
☐ docs.google.com		✓
☐ drive.google.com		✓
☐ amazonaws.com		✓

Advanced

Submit indexing and removal requests via Google Search Console. Learn more

FIGURE 5.2   Creating and configuring a new CSE, step 2 – adding sites to search.

Name the CSE *Document Storage.*

Now go to the Public URL that you can find in the Control Panel to try your new CSE (Figure 5.3).

View it on the web | Public URL |

FIGURE 5.3   Creating and configuring a new CSE – finding the public URL.

Here is an example search executed on our version of this CSE: orgchart CEO CTO CFO.

(You can do the same search on Google using a long "OR" of the *site:* operators – e.g., *site:amazonaws.com OR site:drive.google.com OR....* However, you would need to paste the substring with "site:"s every time you search, and you may also run into search string limits. With the CSE, you avoid either inconvenience.)

"SOFT" SITE SEARCH

Let's talk in more detail about the "Search the entire web" option.

By default, a CSE will only find results from the sites under "Sites to search." However, one useful technique is to expand the results to all websites, preferring the ones listed. We call this "soft X-Ray." To open your search results up to the whole web, turn on the "Search the entire web" feature, as shown below:

"Emphasizing included sites" is unavailable in regular Google.

Compare the standard (left) vs. "soft" (right) site inclusion (Figure 5.4):

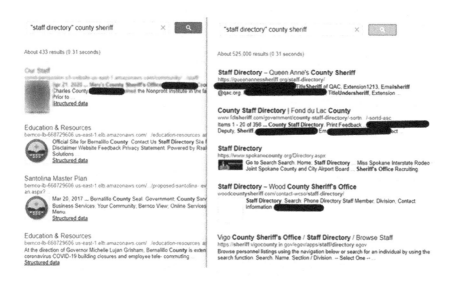

FIGURE 5.4 Comparison of search results for a CSE with no included sites vs. results with the new CSE's based on a list of document storage sites.

You can get even finer control over sites' inclusion and exclusion through XML configuration files. We wrote about them in a dedicated section.

URL PATTERNS

You can include specific sets of pages to search (vs. whole sites) in a CSE via URL *patterns* that use the asterisk (*).

Google, too, can search for sites with an asterisk – for example, site:behance.net/*/resume. However, Google's pattern site search often misses lots of results.

To try using a pattern, press "Add Site" and enter this expression (Figure 5.5):

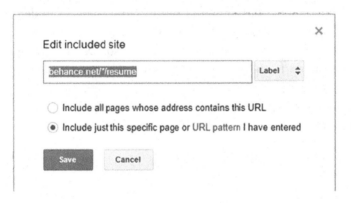

FIGURE 5.5   Creating and configuring a new CSE – using URL patterns.

The resulting CSE will look for pages in the format *https://behance. com/<anythign here>/resume.*

You can use the asterisk * to search through all indexed pages from *.com* sites in this fashion (Figure 5.6):

Sites to search

Add	Delete	Filter	Label ▾	1- 1 of 1 ‹ ›
☐ Site		Label		Available in Site Restricted
☐ *.com			This pattern is not specific enough or searches too many domains.	

FIGURE 5.6   Another URL pattern example: search all .com websites.

(Do not worry about the above "pattern is not specific" warning and a red cross icon that you may see. The CSE will work correctly. You can include *.com, *.edu, *.org, *.uk, etc.)

A pattern ending in /*resume will look for the end of the URL to contain the word *resume* (or any other keyword you specify).

Here is an example search. (Note that throughout the book, underlined "CSE" section titles point to various Custom Search Engines, which you can try by following the link. We encourage you to recreate them.)

Careers CSE

The Careers CSE (Figure 5.7), with the patterns *.com/*career and *.com/*job and no sites included, will find URLs ending in *jobs* and *careers*.

Example searches:

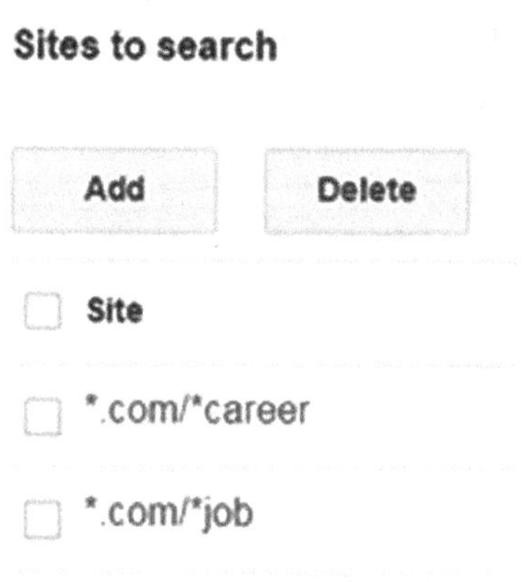

Sites to search

Add   Delete

☐ **Site**

☐ *.com/*career

☐ *.com/*job

FIGURE 5.7   URL patterns for duplicating the Careers CSE.

- devops

- senior software director

- rn houston

Behance Resumes CSE

Behance is a social media platform owned by Adobe, which aims "to showcase and discover creative work." Behance was founded by Matias Corea and Scott Belsky in November 2005. The company was acquired by Adobe

Systems for $150 million in December 2012. In July 2018, Behance had over 10 million members (*source:* Wikipedia).

The Behance Resumes CSE adds a URL pattern *(behance.net/*/resume)* in the "sites to include" setting.

Examples:

- UI designer Boston

- Illustrator

- Award-winning

"KEYWORDS"

The "Keywords" (Figure 5.8) in the CSE setup are optional. When you fill them out, it is not that they will be automatically added to the user's queries. Instead, Keywords increase the ranking of pages with the entered terms but do not drop other pages from search results. This capability is unavailable on Google.

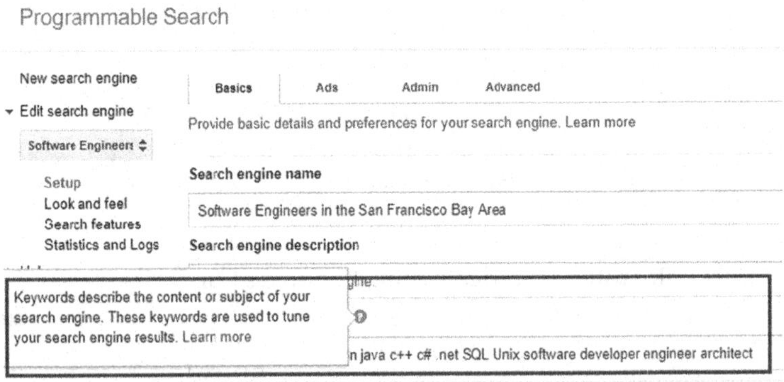

Programmable Search

New search engine | **Basics** | Ads | Admin | Advanced
▾ Edit search engine
Software Engineer ⇕

Provide basic details and preferences for your search engine. Learn more

Setup
Look and feel · Search engine name
Search features
Statistics and Logs · Software Engineers in the San Francisco Bay Area

Search engine description

Keywords describe the content or subject of your search engine. These keywords are used to tune your search engine results. Learn more

n java c++ c# .net SQL Unix software developer engineer architect

FIGURE 5.8   CSE control panel featuring the "Keywords" setting field.

As an example, try the COVID CSE. We have included one term, *COVID-19,* in the keywords and had the CSE Search the entire web.

Example search: unemployment (Figure 5.9)

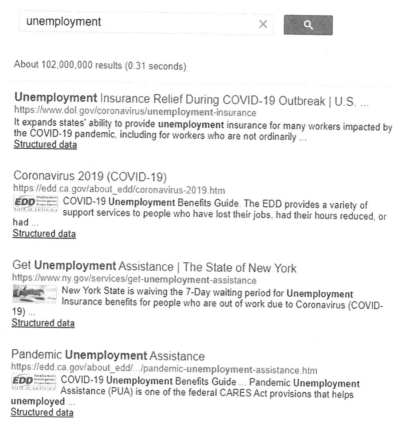

FIGURE 5.9 Sample search and results for the COVID-19 CSE.

AUTOMATICALLY ADD SEARCH TERMS

Use Search features (on the right)/Advanced (on the top) to enter a search string to be automatically – and invisibly – appended to the end-user's search. You need to save the setting for it to take effect.

PDF Files CSE

Running any search in the PDF Files CSE, you can see that the results are exclusively PDF files in this CSE (Figure 5.10):

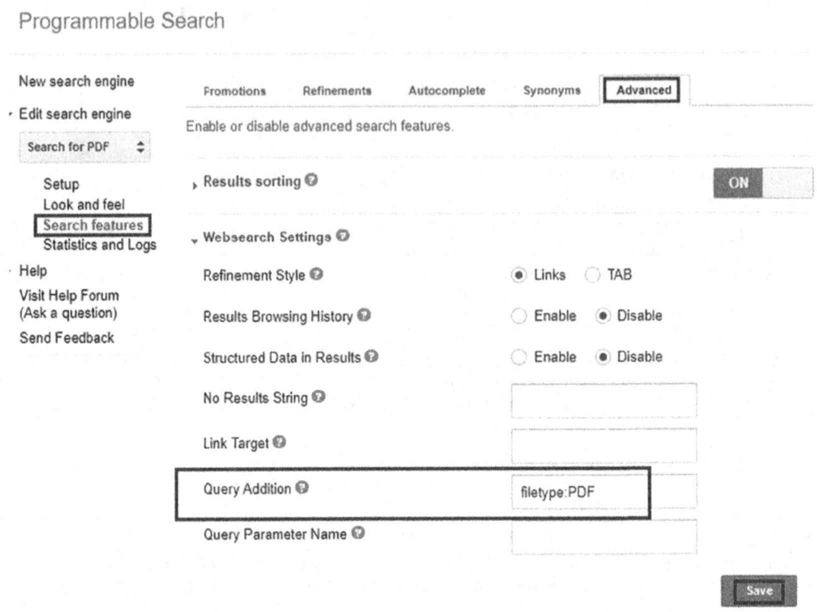

Programmable Search

| | Promotions | Refinements | Autocomplete | Synonyms | Advanced |

New search engine

· Edit search engine

Search for PDF ⇕

Enable or disable advanced search features.

Setup
Look and feel
Search features
Statistics and Logs

Help
Visit Help Forum
(Ask a question)
Send Feedback

, Results sorting ❓   ON

⌄ Websearch Settings ❓

Refinement Style ❓   ● Links   ○ TAB

Results Browsing History ❓   ○ Enable   ● Disable

Structured Data in Results ❓   ○ Enable   ● Disable

No Results String ❓

Link Target ❓

Query Addition ❓   filetype:PDF

Query Parameter Name ❓

Save

FIGURE 5.10 CSE control panel showing the Advanced tab of the Search features screen, highlighting the Query Addition field under Websearch settings for automatically adding terms to a search.

- resume engineer java mongodb amsterdam

Let's build another CSE that searches the whole web with a simple auto-addition.

Gmails CSE

For the Gmails CSE, *"gmails.com"* is automatically added to every search. It will help to discover Gmail-based addresses.

Example search (Figure 5.11):

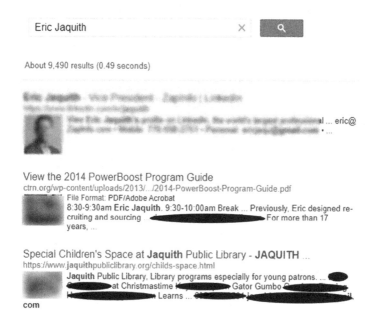

FIGURE 5.11   Sample search results for the Gmails CSE, demonstrating the effect of automatically including "gmail.com" in every search.

CHAPTER 6

Search Refinements

R EFINEMENTS ARE "LABELED" PREDEFINED search strings, added to the end-user's search. The term "label" means a site or set of sites in which the search is performed.

For the end-user, Refinements look like clickable tabs under the search box. They allow narrowing the search down in various ways.

An example use of Refinements is the <u>Documents by Format CSE</u>.

In this CSE, the end-user can enter terms and then narrow down to a selected document format, such as Word, text, and PDF (Figure 6.1).

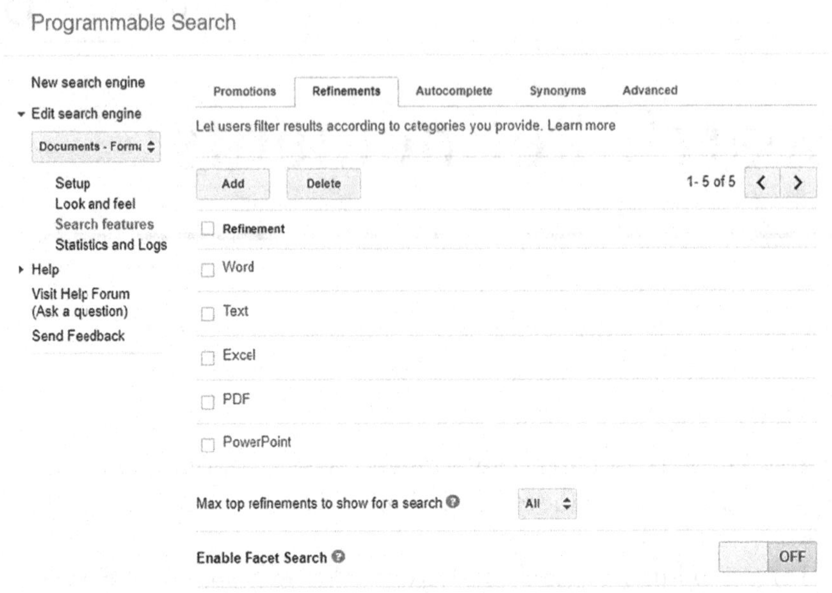

Programmable Search

| | Promotions | **Refinements** | Autocomplete | Synonyms | Advanced |

New search engine

▾ Edit search engine

Documents - Form: ⬍

Let users filter results according to categories you provide. Learn more

Setup
Look and feel
Search features
Statistics and Logs
▸ Help
Visit Help Forum
(Ask a question)
Send Feedback

| Add | Delete | | 1- 5 of 5 ⟨ ⟩ |

☐ **Refinement**

☐ Word

☐ Text

☐ Excel

☐ PDF

☐ PowerPoint

Max top refinements to show for a search ❓   All ⬍

Enable Facet Search ❓   OFF

FIGURE 6.1 Control Panel showing the Refinements tab of the Search features screen with several Refinements added to a CSE.

To start adding refinements to your CSE, go to Control Panel/Advanced (on the left)/Refinements (on the top). Click the "Add" button and then define what each Refinement stands for and save.

We have defined the Refinement "Word" as *filetype:doc OR filetype:docx* (and other refinements, similarly) (Figure 6.2):

FIGURE 6.2   CSE Control Panel showing the Refinements tab of the Search features screen, with the Add Refinement interface open to demonstrate defining a refinement.

(We will explain Knowledge Graph Entities in a relevant section below.) Example searches:

- how to build custom search engines
- resume book 2020
- competitive intelligence

CHAPTER 7

Synonyms

GOOGLE WILL AUTOMATICALLY SEARCH for keyword synonyms and variations; it is its built-in feature. But if you want to identify related words that may not quite be considered synonyms, this CSE mechanism accommodates that.

Of course, you can define legitimate synonyms (e.g., set *curriculum vitae* and *resume* as synonyms of *CV*). But nobody will check whether the synonyms you enter are "correct" in English. You may want to play with the setting, defining words with different meanings as synonyms, and see what happens.

We have discovered that the Synonyms feature allows running queries equal to very long "OR" Boolean statements. A practical implementation is to enter a term in Synonyms that means a category, define its synonyms as objects of the type, and use the term as a shortcut when searching. For example, you could define a list of (up to ten) companies as synonyms of the term *mycompanies*.

Searching with long OR statements is especially useful in diversity sourcing. For example, you may search for women's first names, Latino last names, or alumni of diversity colleges. Another use case is searching for target companies or locations. Or, search for Ivy League schools or Fortune 500 companies.

To define synonyms for your CSE, you can either use the Control Panel or create an XML file with synonym definitions. For large lists of synonyms, the file-based method is more convenient. We will explain the process for the example below. Refer to Google's <u>documentation</u> for additional details about synonyms.

The synonyms limit is 500 terms and ten synonyms for each term, with up to 2K total variants.

SYNONYMS EXAMPLE

Let's build a CSE with Synonyms.

Create a CSE to search for LinkedIn profiles. Go to the Control Panel/ Advanced/Synonyms. Start adding a synonym:

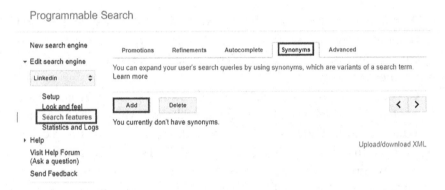

FIGURE 7.1   CSE Control Panel showing the Synonyms tab of the Search features screen, highlighting the Add synonym button.

When you press "Add," you will see a box for adding synonyms, up to ten for each term:

FIGURE 7.2   Continuation from Figure 6.2, showing the interface for adding a new synonym.

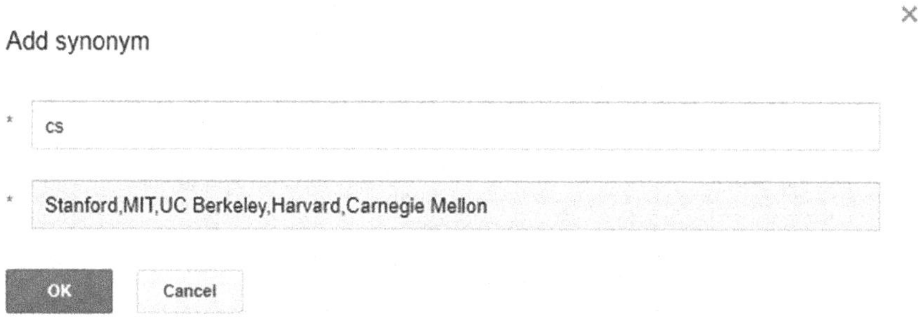

×

Add synonym

cs

Stanford,MIT,UC Berkeley,Harvard,Carnegie Mellon

OK   Cancel

FIGURE 7.3   Continuation from Figure 7.1, showing a sample synonym definition.

Let us try this synonym:

In the resulting CSE, entering *CS* would find not only the keyword *CS* but also *Harvard, Stanford, MIT, Berkeley*, and *Carnegie Mellon*. You will not necessarily see the school names in snippets, but the found pages will have them (Figures 7.2 and 7.3).

Synonyms Example CSE #1

Here is an example search to demonstrate how the <u>Synonyms Example CSE #1</u> works:

- <u>cs chief software architect seattle</u>

You can search for *cs -cs*, thus no longer finding *cs* as a keyword but only finding its synonyms, which are school names:

- <u>cs -cs CEO</u>

(Note, by the way, that on Google, you can also search for *<keyword> -<keyword>* to include *true* synonyms and variations for the *<keyword>* but not the *<keyword>* itself. Example: <u>recruiter -recruiter</u>).

Create a CSE with this synonym and try a few searches. Create a CSE with another set of synonyms. For example, create a CSE with synonyms for *sales* as *account executive, account manager*, and *business development* and try it out.

You can add several synonyms, along with their values, in the Control Panel.

WORKING WITH SYNONYMS XML FILES

In the Control Panel, you cannot see all of the synonyms' definitions at once, and it takes time to type in new ones. It makes editing inconvenient. It is more practical to edit the Synonyms XML file.

To get a sample XML-formatted file, do the following. Define two synonyms, for example, these:

☐ Search Term	Synonyms
☐ CS	MIT,Stanford,UC Berkeley
☐ CS1	Harvard,Carnegie Mellon

FIGURE 7.4 Example synonym definitions within a CSE.

Synonyms Example CSE #2

The search cs OR cs1 will produce the same results in the Synonyms Example CSE #2 as the previous CSE does.

Recreate the CSE and export the XML file. The file will look like this (Figures 7.5 and 7.6):

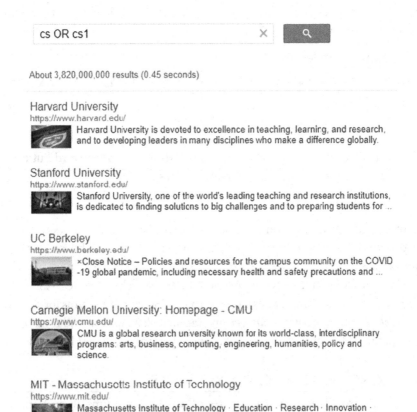

FIGURE 7.5 Example search results demonstrating the impact of the two synonyms defined in Figure 7.4.

```
<?xml version="1.0" encoding="UTF-8"?>
<Synonyms start="0" num="2" total="2">
  <Synonym term="CS">
    <Variant>MIT</Variant>
    <Variant>Stanford</Variant>
    <Variant>UC Berkeley</Variant>
  </Synonym>
  <Synonym term="CS1">
    <Variant>Harvard</Variant>
    <Variant>Carnegie Mellon</Variant>
  </Synonym>
</Synonyms>
```

FIGURE 7.6   Sample of the XML code produced by exporting the synonyms.xml definitions file from a CSE with the definitions shown in Figure 7.4.

The XML format is clear. You can edit the file, adding more synonyms (we recommend using Notepad++), and then upload it back using the Control Panel.

HOW TO USE SYNONYMS TO IMPLEMENT LONG OR STATEMENTS

Here is a "hack" we have discovered. It helps to overcome Google's limit of 32 keywords per search.

You can arrange to auto-add a search string with many ORs by doing the following:

1. Define *f1* to have ten synonyms from the list of your terms to search, *f2* the next ten words, etc. (We have picked *f1, f2,* etc., just as random terms.)

2. Set the CSE to append this string automatically: *f1 OR f2 OR f3 OR f4 OR f5 OR f6 OR f7 OR f8 OR f9 OR f10*

As this string is added, it creates a logical OR for *a hundred* terms (in this case). Using this technique, you can push the limit of keywords *from Google's 32 to 320!*

The Fortune 100 Companies CSE implements a search for the names of all Fortune 100 companies. This CSE combines the synonyms and auto-append features to add a long OR of the Fortune 100 company names to every search. The Synonyms XML file looks like this (Figure 7.7):

```
<?xml version="1.0" encoding="UTF-8"?>
- <Synonyms total="10" num="10" start="0">
  - <Synonym term="f1">
        <Variant>Walmart</Variant>
        <Variant>Exxon Mobil</Variant>
        <Variant>Apple</Variant>
        <Variant>Berkshire Hathaway</Variant>
        <Variant>Amazon.com</Variant>
        <Variant>UnitedHealth Group</Variant>
        <Variant>McKesson</Variant>
        <Variant>CVS Health</Variant>
        <Variant>AT T</Variant>
        <Variant>AmerisourceBergen</Variant>
    </Synonym>
  - <Synonym term="f2">
        <Variant>Chevron</Variant>
        <Variant>Ford Motor</Variant>
        <Variant>General Motors</Variant>
        <Variant>Costco Wholesale</Variant>
        <Variant>Alphabet</Variant>
        <Variant>Cardinal Health</Variant>
        <Variant>Walgreens Boots Alliance</Variant>
        <Variant>JPMorgan Chase</Variant>
        <Variant>Verizon Communications</Variant>
        <Variant>Kroger</Variant>
    </Synonym>
  - <Synonym term="f3">
        <Variant>General Electric</Variant>
        <Variant>Fannie Mae</Variant>
        <Variant>Phillips 66</Variant>
        <Variant>Valero Energy</Variant>
        <Variant>Bank of America</Variant>
        <Variant>Microsoft</Variant>
        <Variant>Home Depot</Variant>
        <Variant>Boeing</Variant>
        <Variant>Wells Fargo</Variant>
        <Variant>Citigroup</Variant>
    </Synonym>
  - <Synonym term="f4">
        <Variant>Marathon Petroleum</Variant>
        <Variant>Comcast</Variant>
        <Variant>Anthem</Variant>
        <Variant>Dell Technologies</Variant>
        <Variant>DuPont de Nemours</Variant>
```

FIGURE 7.7 Sample of the XML code from a synonyms.xml file that, when imported, will add ten new synonyms to a CSE, each representing the names of ten of the Fortune 100 companies.

Example searches:

- <u>layoffs</u> – will bring a mix of results about layoffs

- <u>appointed ceo</u> – news about CEO appointments

- <u>earnings for the second quarter</u> – news about company earnings for Q2

Women's Names (LinkedIn) CSE

The <u>Women's Names (LinkedIn) CSE</u> auto-appends a long OR of common US women names to the end-user search. It takes advantage of the Synonyms feature and is similar to the Fortune 100 CSE. The Synonyms XML file looks like this (Figure 7.8):

```xml
<?xml version="1.0" encoding="UTF-8"?>
<Synonyms total="5" num="5" start="0">
 - <Synonym term="f1">
        <Variant>Emma</Variant>
        <Variant>Olivia</Variant>
        <Variant>Ava</Variant>
        <Variant>Isabella</Variant>
        <Variant>Sophia</Variant>
        <Variant>Charlotte</Variant>
        <Variant>Mia</Variant>
        <Variant>Amelia</Variant>
        <Variant>Harper</Variant>
        <Variant>Evelyn</Variant>
    </Synonym>
 - <Synonym term="f2">
        <Variant>Abigail</Variant>
        <Variant>Emily</Variant>
        <Variant>Elizabeth</Variant>
        <Variant>Mila</Variant>
        <Variant>Ella</Variant>
        <Variant>Avery</Variant>
        <Variant>Sofia</Variant>
        <Variant>Camila</Variant>
        <Variant>Aria</Variant>
        <Variant>Scarlett</Variant>
    </Synonym>
 - <Synonym term="f3">
        <Variant>Victoria</Variant>
        <Variant>Madison</Variant>
        <Variant>Luna</Variant>
        <Variant>Grace</Variant>
        <Variant>Chloe</Variant>
        <Variant>Penelope</Variant>
        <Variant>Layla</Variant>
        <Variant>Riley</Variant>
        <Variant>Zoey</Variant>
        <Variant>Nora</Variant>
    </Synonym>
    <Synonym term="f4">
        <Variant>Lily</Variant>
        <Variant>Eleanor</Variant>
        <Variant>Hannah</Variant>
        <Variant>Lillian</Variant>
        <Variant>Addison</Variant>
        <Variant>Aubrey</Variant>
```

FIGURE 7.8 Sample of the XML code from a synonyms.xml file that, when imported, will add five new synonyms to a CSE, each representing ten of the most popular names for women in the US.

Example searches:

- <u>ios developer bay area</u> – will find women's LinkedIn profiles containing the keywords ios and developer
- <u>vice president HR</u> – profiles of vice presidents of human resources
- <u>plant manager</u> – profiles with the words *plant* and *manager*

Another use of Synonyms is to create abbreviations (shortcuts) of your most commonly used words and phrases.

CHAPTER 8

Configuration Files

To get even finer control over CSE search results, instead of editing your CSE via the Control Panel, you can export, edit and import back two <u>XML-formatted</u> *configuration files* that define the CSE. The files are called *CSE context* and *CSE annotations*. (There's no good reason to have two separate files; we have them for historical reasons.)

You can edit the configuration files to set extra control over the "soft" inclusion of sites and several other parameters. You do not need to know a lot about formatting XML files; editing is straightforward. (But you do need to understand what each field means.)

After creating a simple CSE, you can download the XML files from the Control Panel, edit the files, and then import them back (Figure 8.1).

The two files combined (along with the Synonyms XML if necessary) define what your CSE does. Through downloading, editing, and importing configuration files, you can change the CSE parameters.

Uploading XML context and annotation files allow you to set some parameters in ways beyond those available in the Control Panel. For example, you will be able not only to include ("FILTER") or exclude ("ELIMINATE") a site from search but also assign different sites different numerical weights between −1 and 1 ("BOOST"). (*FILTER, ELIMINATE,*

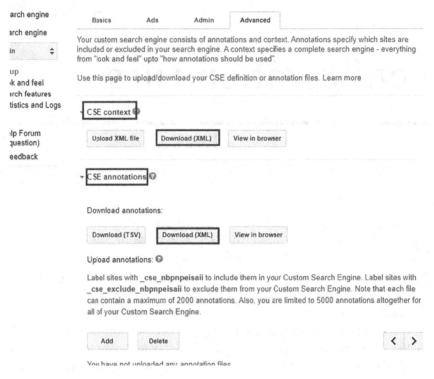

FIGURE 8.1   The CSE Control Panel showing the Advanced tab of the Setup screen, highlighting the buttons to download a CSE's context.xml and annotations.xml definition files.

and *BOOST* are the values you can assign to various labels – i.e., the sites to search.)

Here is how to do it. Download the XML context file. It will have a piece marked "BackGroundLabels" that looks like this (Figure 8.2):

```
- <BackgroundLabels>
    <Label mode="BOOST" name="_cse_xcvjlqko6es"/>
    <Label mode="ELIMINATE" name="_cse_exclude_xcvjlqko6es"/>
  </BackgroundLabels>
```

FIGURE 8.2   Sample XML code from a CSE's context.xml definitions file, demonstrating the options to boost a specific site's pages in the search results or eliminate that specific site from the results.

Here, you can add and edit labels using a format that includes weights (Figure 8.3):

```
<BackgroundLabels>
    <Label name="_cse_5mra7n2kfcu" mode="BOOST" weight="0.4"/>
    <Label name="_cse_exclude_5mra7n2kfcu" mode="ELIMINATE"/>
</BackgroundLabels>
```

FIGURE 8.3 Sample XML code from a CSE's context.xml definitions file, demonstrating how to edit the XML from Figure 8.1 to add a specific weight to boost a specific site's pages in the search results.

"BOOST" without a weight specified gets a weight of 0.7 (whatever that means).

(Note that in XML format, the order of lines does not matter as long as the format follows the correct syntax.)

Notably, you have a choice for a less-desired site to be either excluded ("ELIMINATE") or only included when it is a great match (equivalent to "BOOST" with a weight of -1).

You can read more at Creating Custom Search Engine with configuration files, and specifically about configuration files and context files, in Google's help.

BACKING UP, SHARING, AND DUPLICATING

Whether you will edit the configuration files or not, saving both files allows you to create a CSE copy or revert to a previous CSE version. You can regularly save the files as a way to keep your CSE version history.

Sharing configuration files is a way to give someone else access to all the CSE parameter settings, with clearly identified custom values down to weights for included sites (if you set those).

Another option to share a CSE is to add collaborators via the CSE Control Panel, but the invited person will need to click many links to learn about the CSE parameters.

Other than that, you can explain which parameters you have set to which values in a human language so that others can reproduce your CSE. That is what we do in the book. (Neither of the sharing options is convenient, but we have said that the CSE UI was quite outdated.)

If you were wondering, it is nearly impossible to reverse-engineer or duplicate a CSE given only its URL.

CHAPTER 9

Other Features of Note

LOCALIZATION

You can restrict CSE results to a country or language in the Control Panel settings. You can do so on Google by using the <u>Advanced Search Dialog</u>, but with a CSE, you can set the values that will be "on" for all searches.

You will find the country and language settings under two different tabs in the Control Panel (Figures 9.1, 9.2, and 9.3):

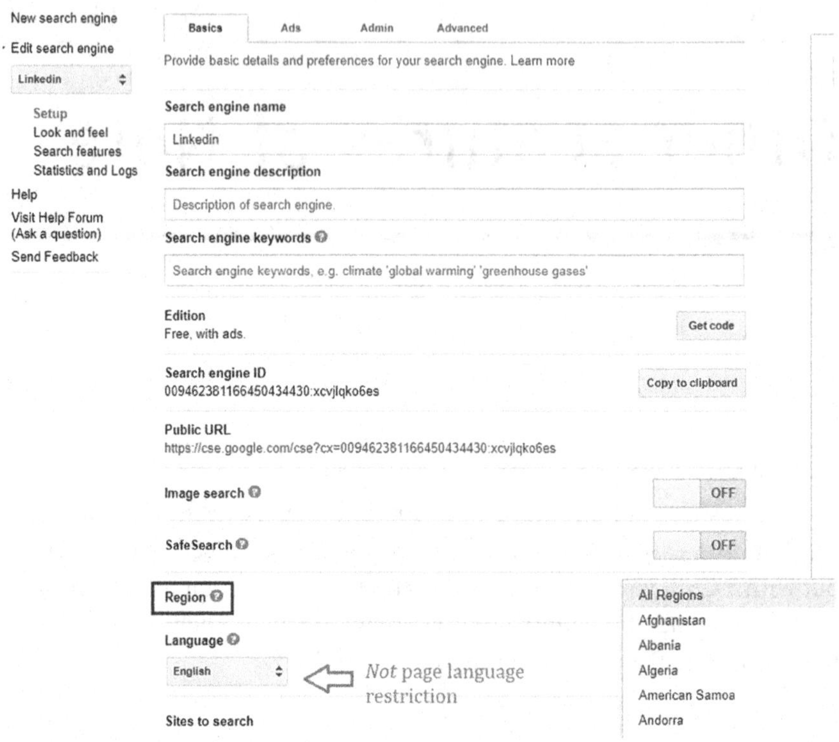

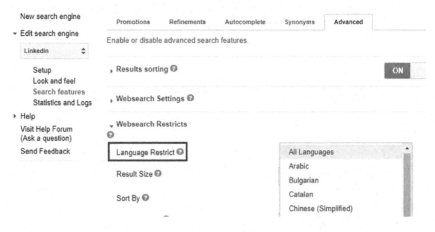

FIGURE 9.1   The CSE Control Panel, showing the Basics tab of the Setup screen, highlighting how to set a CSE to look for results from a specific country.

FIGURE 9.2   The CSE Control Panel showing the Advanced tab of the Search features screen, highlighting how to set a CSE to look for results in a specific language.

Additionally, you have an option to *boost* results by country (unavailable in Google).

FIGURE 9.3 Addendum to Figure 9.1, when a country is selected, the option to restrict results to the country is revealed. When it is turned off, results from the selected country are *boosted;* turned on, all results will be from that country only.

IMAGE SEARCH

You can set the Image search "on" in the Control Panel. As a result, your CSE will get a *Web* tab and an additional *Image* tab. If you wish, you can make Image search the default.

Image search has its separate settings, including a query addition and even language. Like the CSE Web search, CSE Image search returns no more than 100 results per query.

In our experience, Image search often misses results compared to Google's image search. We have only seen results for wide-open searches.

Search Images CSE

The Search Images CSE does not include any sites and has no other parameters. It has an Image search tab.

Example searches for images:

- meet our CFO (see a screenshot below)

- top companies in oil and gas

- accountant certification in australia

Go back to the "Google Search" chapter of the book and run some of the sample searches we have shared. Notice the difference in the results (Figure 9.4).

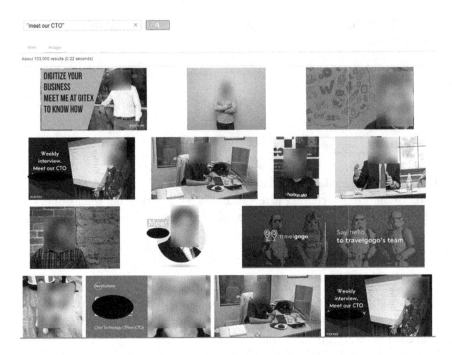

FIGURE 9.4   Sample search results for the Search Images CSE.

As in Google's Image search, you can use advanced Google operators, such as *site:*, *inurl:*, *intitle:*, and the image-specific operator *imagesize:*.

As an example application, since LinkedIn profile photos have the size 200 x 200 pixels, you can find them (and not other images) as follows (Figure 9.5):

- site:linkedin.com/in imagesize:200x200

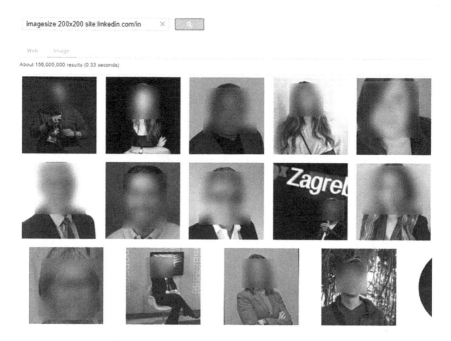

FIGURE 9.5   Sample search results for the Search Images CSE, demonstrating the *imagesize:* operator.

Unfortunately, CSE-specific operators do not work in the Image tab. There is also no way to filter by image type (such as "photo" or "face") or color.

Compared to Google's image search, you will see a more compact display since all images are scaled to the same height.

It might be useful to include the image search option in your CSEs as a visual way to search with the filters that CSEs provide.

CHAPTER 10

Troubleshoot Your CSE

Y OU HAVE CREATED A CSE, but it doesn't search in the ways you expect. Here are some things you can look at:

1. Does your CSE return results on simple searches like *Microsoft* or *partner*? If it does not, it might be best to start over.

2. Does your CSE have many settings? You may want to remove some and see how its behavior is affected. (It is hard to look at many parameters at once.) Create a CSE with one or two settings that you want to use and see if it works in the ways you expect. Or recreate your CSE, testing it after you add each parameter.

3. Search with a simple keyword – for example, *manager.* Do the results make sense? Are they coming from the sites you have specified?

4. Have you selected the option "Search the entire web"? You will then see results outside of the sites you have specified, which may not be your intention.

5. Check the refinements for the option "Search the entire web."

6. Download your configuration XML files from the "advanced" panel and look through them. The files are easier to review than each of the settings in the Control Panel (use Notepad++).

3

Discover *more:* Advanced CSEs

In this part, we will tell you about the groundbreaking CSE functionality allowing you to search parts of the web as if they were structured databases with various types of objects and values.

Structured web search functionality – i.e., searching the web using filters (such as *Job Title = Senior Sustainability Manager* and *Location = New York*) – is available both during the CSE creation stage and for end-users. To understand how it works, you need to learn a little about Schema.org (and other standard types of) objects represented by web pages.

CHAPTER 11

Metadata Types

SCHEMA.ORG OBJECTS

FIGURE 11.1   The Schema.org logo.

Schema.org is a standard set of types of objects intended for webmasters to include structured data within pages invisibly and for search engines to notice (Figure 11.1).

Objects, such as *Person*, along with values, such as *Jobtitle*, can be embedded in the source code by the page creator. Googlebot recognizes the objects and values and stores this information in its Index, along with the page content.

The schemas are a set of objects, each of which is, in turn, associated with a set of properties. The types form a hierarchy. The Schema.org vocabulary currently consists of 829 Types, 1,351 Properties, and 339 Enumeration values (*source:* https://schema.org/docs/schemas.html).

SCHEMA.ORG OBJECTS AND CUSTOM SNIPPETS IN GOOGLE SEARCH

Google uses embedded structured data to provide *rich snippets* in search results. We have previously defined a snippet as the preview of a page's contents shown on the search results page (SERP). Regular snippets contain only text, but rich snippets allow Google to provide a wide range of additional information. When you see, for instance, movie or product star ratings (Figure 11.2):

www.imdb.com › title

"Man Down" Diversity (TV Episode 2015) - IMDb

Alarmed by his lack of political correctness, headmistress Miss Lipsey sends Dan on a **diversity** awareness course. Meanwhile, at the beauty salon Jo comes up ...

★★★★★ Rating: 7.8/10 - 51 votes

www.imdb.com › title

"The Office" Diversity Day (TV Episode 2005) - IMDb

Directed by Ken Kwapis. With Steve Carell, Rainn Wilson, John Krasinski, Jenna Fischer. Michael's off color remark puts a sensitivity trainer in the office for a ...

★★★★★ Rating: 8.3/10 - 4,759 votes

www.imdb.com › title › characters

"Man Down" Diversity (TV Episode 2015) - Stephanie Cole as ...

"Man Down" **Diversity** (TV Episode 2015) Stephanie Cole as Nesta.

www.imdb.com › title

"Mostly 4 Millennials" Diversity (TV Episode 2018) - IMDb

Directed by Derrick Beckles. With Meghan Ryan, Derrick Beckles, Fred Durst, Roger Hervas. Millennial hero Derrick Beckles interviews millennial reality star ...

★★★★★ Rating: 6.7/10 - 9 votes

FIGURE 11.2 Sample Google search results showing how Google incorporates Schema.org data into the results.

Googlebot recognizes and saves various objects and values in the Index, along with the rest of the page's content. For example, here is some of the metadata (from Schema.org objects) for those movie reviews (Figure 11.3):

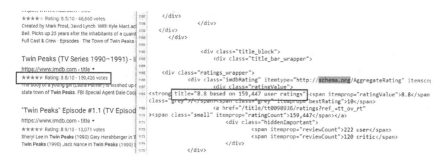

FIGURE 11.3   Google search results side by side with the HTML code for those results, highlighting that the data is from a Schema.org object.

MICROFORMATS

Microformats define an alternative to Schema.org's way of incorporating structured data within a web page.

The most common Microformats object in use is hCard. hCard has fields for a person, including name, company, phone, and address. Sometimes the same information as in *hCard* is duplicated in a Schema. org *Person* object, but *hCard* often stores unique values, and you can search for them.

Another Microformat of interest is hCalendar. It represents an event, much as hCard represents a person. hCalendar has fields for start and end dates, a summary or description, URL, and location, among others. As with hCard, Schema.org offers an alternative in the Event object, but you can search for each possible data field separately.

META TAGS

Another type of structured data is included in web pages within <meta> HTML tags. For some sites, Meta tags include values of interest that other structures do not. For example, the bio you see on a GitHub profile is also a field in the page's Meta tags and is not present elsewhere. *<meta>* is a type of HTML tag containing information that guides User Agents (web browsers and web crawling programs like Googlebot) in the correct way to display and process the rest of the code that makes up the web page. An example use of Meta tags is adjusting the way text and images are laid out on a page depending on the size of the screen or window.

But within this book, we are most interested in the application to search engines. Unlike Schema.org and Microformats, Meta tags have no formal standard for the content and structure. However, as most webmasters want their pages to be correctly indexed and represented by Google and other search engines, they will follow best practices defined by the search engine industry.

CHAPTER 12

Schema.org and Custom Search

W HILE RICH SNIPPETS AND additional details in search results provide value to everyone, what is of utmost interest to us is that CSEs can also *search* for metadata. In fact, one can either search for the existence of a particular Schema.org object or for objects on pages that contain specific fields or values. For example, using a CSE, one can search for a person with a given job title or employer! Google does not have that remarkable ability.

BUILDING A CSE WITH A SCHEMA.ORG OBJECT

The simplest way to use Schema.org objects in your CSE is to add them via the Control Panel. The option to restrict search results so that only pages containing a given object are listed is at the bottom of the Settings -> Basic tab (Figure 12.1).

To select an object, such as Person, start typing its name in the text box and then choose it from the suggestions displayed below. The CSE pictured above will now only return pages with a Person object. You can add up to ten objects this way, per CSE. Adding multiple Schema.org objects is like adding several URLs or URL patterns to the Sites to Search or Sites to Exclude: the CSE will look for pages that contain at least one of the defined objects. However, we have found that limiting CSEs to a single object (and creating a separate CSE for the next object you want to look for) leads to more productive outcomes.

Restrict Pages using Schema.org Types ⓘ

Restrict pages from the above site list to only those that contain Schema.org types from the list below.

You can add up to ten (10) schema.org types to your Search Engine. Note that when you add a node, all its children automatically get included, so you do not need to add them again. For example, if you add CreativeWork, you do not need to add Book, ImageObject, VideoObject etc. separately.

FIGURE 12.1   The CSE Control Panel, showing the bottom of the Basics tab of the Setup screen, demonstrating how to add a Schema.org object search restriction to a CSE.

TIP: EASILY BUILD CSEs FOR PUBLIC PROFILES

A common use case when searching for people or information about people, such as in recruitment, is to X-Ray search social websites for user profiles. While you can X-Ray many social sites from Google, using the site: operator, it is often tricky to find a search string that produces users' profiles in the results. With CSEs, if those profiles include the Schema.org Person object, you can make that tricky process trivial with just two steps. First, set your target website under Sites to search. Then add Person under Restrict pages by Schema.org objects. Now you have a powerful X-Ray CSE for that site! Every result from the CSE will be a user profile from your target social site.

For example, the <u>Psychology Today CSE</u> includes only one site to search: <u>psychologytoday.com</u>. The Psychology Today website contains articles about psychology as well as directories of therapists, and individual therapist profiles. The *Person* object restricts results. Run these example searches:

- <u>LCP illinois</u>

- <u>school psychologist new york</u>

- <u>professor of psychology</u>

You will see that the results contain only directories and individual therapist profiles. Now, try developing a search string for regular Google searches that will so precisely restrict the results. That is quite a challenge!

BEYOND THE PERSON

CSEs can restrict to Schema.org object types other than Person. Some interesting to explore objects are the following:

- Physician

- Accountant

- (Software) Code

- Job post

- Blog post

- Organization

- MedicalBusiness

- MedicalTrial

- MedicalSpecialty

- MedicalClinic

- Occupation

- Place

- ProfilePage

- ProfessionalService

- SearchAction

- TVSeason

- Thesis

- Zoo

and more.

The <u>HowToTool CSE</u> has no sites to search or exclude. The results are pages with the *HowToTool* object. This CSE will find advice on which tools to use to solve any problem! Some example searches include the following:

- <u>SEO</u>

- <u>Python</u>

- <u>cake</u>

- <u>exercise</u>

- <u>meaning of life</u>

(Create a CSE with the *HowTo* object. With it, you will find mostly cake recipes, which would be hard to guess based on the object name. Once again, the example shows that webmasters exercise a good deal of freedom in applying Schema.org's structure to their needs.)

CHAPTER 13

Knowledge Graphs

B EFORE DIVING DEEPER INTO CSEs' ability to search for the different metadata objects described above, it is worth noting that there is one other, very different, type of metadata: the Knowledge Graph Entity (KGE).

Google announced Knowledge Graph (KG) on May 16, 2012, as a way to significantly enhance the value of information returned by Google searches (*sources:* Wikipedia; blog post Introducing the Knowledge Graph: things, not strings). Google used Wikipedia, CIA World Factbook and Freebase "to gather data about people, events, animals…history and other topics" (source: blog post The Beginner's Guide to Google's Knowledge Graph). Even though Wikipedia gets less traffic due to the Knowledge Graph object displays, the company welcomes collaboration with Google.

Google keeps and expands the KG. Each object in the KG can be a child or a parent of another object in the Graph.

Here is a visual representation of a small section of the KG (Figure 13.1):

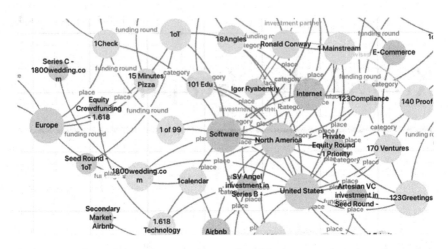

FIGURE 13.1 Visual depiction of a KG model showing the interrelationship between various terms and concepts.

You can find other drawings by <u>Searching in Images</u>.

Google displays the KG object on the right when it identifies the object (i.e., the purpose) a user query implies (Figure 13.2):

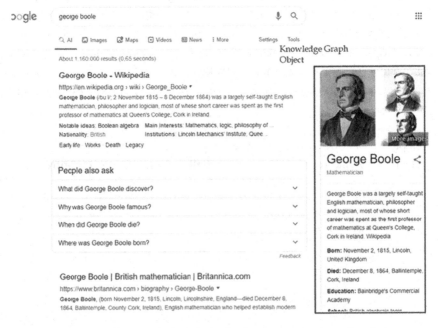

FIGURE 13.2 Sample Google search results demonstrating how Google incorporates KGEs into the search results page.

SELECTING KGEs IN CSEs

CSEs recently got a new setting in the Control Panel that takes <u>semantic search</u> features even further. Choosing a restriction by KGE works the same way as selecting a restriction by Schema.org object. Type the name of an object into the box and choose the one you want from the suggestions (Figure 13.3).

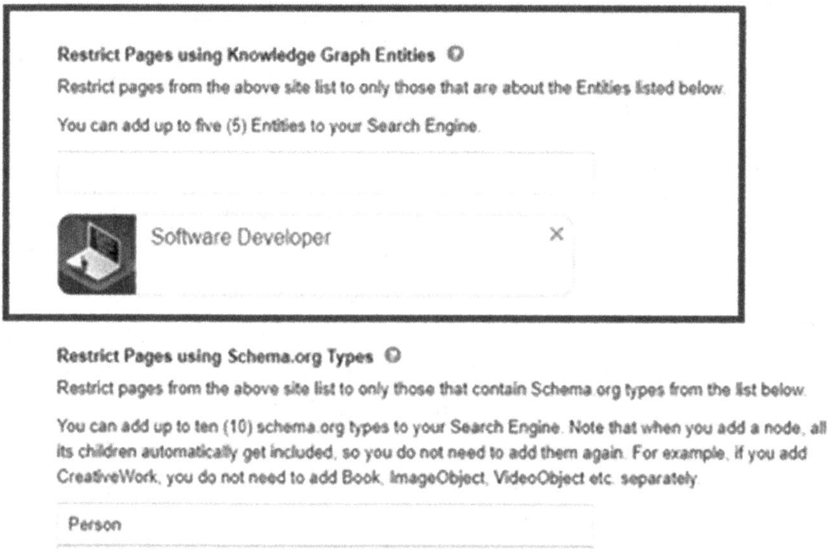

FIGURE 13.3 The CSE Control Panel showing the bottom of the Basics tab of the Setup screen, demonstrating how to add a KGE search restriction to a CSE.

However, in terms of the results your search engine will find, KG restrictions are quite different.

First, while Schema.org objects are stored in metadata in HTML code placed there by web page creators, KG objects do not have representation in HTML, and webmasters cannot control them. Instead, KG objects reflect *Google's* "idea" of which ones are relevant, characteristic for the page. Therefore, the new option covers a much more extensive range of websites that can show up in the results.

Second, Schema.org defines a limited number of standard objects. The number of KG objects is vast and growing. The object list is inaccessible. You can only guess that an entity is an object and then Google to see what

Google presents on the top right. The drop-down object choices in the Control Panel don't list the majority of the available objects. You can Google for publicly shared URLs of objects: "google.com/search?kgmid." But it is difficult to make the search targeted. Additionally, only a small portion of KG objects have made it into Google's Index.

Finally, while Schema.org objects consist of fields with values, KG objects have a *type* (such as "Author" or "City") and no deeper structure.

Restricting results by KG Entities is most useful when you need to gather knowledge on a subject. You will encounter articles, books, and blog posts in the results. We recommend creating several KGE-based CSEs to help gain a feel for how KGEs impact search results.

TIP: KG BOOLEAN

In the Control Panel under Setup -> Basic, you can select up to five KGEs. Just as with the Schema.org objects and the list of sites to search and exclude, this setting entries (KG Objects) form a Boolean OR query. Pages will be relevant to at least one of the defined objects. It is also possible to use a CSE to look for pages that contain both of two different objects (a Boolean AND), though the method is not obvious.

The method uses Refinements (which we discussed in Chapter 6). First, enter one KGE in the Restrict by the Knowledge Graph Entity box under Settings -> Basic. Then navigate to Search features -> Refinements and define a refinement that contains another KGE-based restriction. When an end-user performs a search and selects the refinement, the AND is applied: only results relating to both KGEs will be shown.

EXAMPLES OF CSEs THAT SEARCH FOR KG ENTITIES

The Females Knowledge Graph CSE will bring up content (such as articles and blog posts) that match your other search criteria and which Google has associated with the concept of *female*. Try these example searches and see for yourself:

- careers

- salaries

- work from home

Our Jobs Knowledge Graph CSE shows (mostly) job posts, as in the example java developer san francisco. Performed via the CV Knowledge Graph CSE, the same example of java developer san francisco produces CVs and resumes primarily. The Open Source Intelligence (OSINT) Knowledge Graph CSE will find pages relevant to that topic, such as slides from OSINT presentations. Finally, the Knowledge Graph Test CSE works by defining Refinements to combine KG Entities, as mentioned in the Boolean tip above.

To build this last CSE, we have started with a Search Everything CSE and restricted results to one of two KGEs: results must have to do with either the engineer or the computer programming KGE. Then we added a Refinement that, when selected, applies the woman KGE (Figure 13.4).

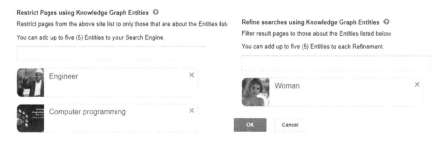

FIGURE 13.4 On the left is the CSE Control Panel showing the bottom of the Basics tab of the Setup screen with the *Engineer* and *Computer programming* KGEs added as restrictions to a CSE. On the right, the same CSE Control Panel, showing the Refinements tab of the Search features screen, with the Woman KGE added as a restriction for that Refinement.

All results from this CSE will have to do with women and either engineering or computer programming. Search for leadership, or conferences, or javascript using the KG Test CSE to see how that works in practice.

CHAPTER 14

Fascinating CSE Advanced Search Operators

A s outlined in Chapter 12, you have already learned that CSEs can search for pages that contain Schema.org objects and that these objects have additional structure in the form of fields and values. In this chapter, we will explain how to search for those fields and values directly and precisely.

For example, a LinkedIn profile page may include a *Person* (object) working at *<Company Name> = value*). With a CSE, using a special operator, you can search for LinkedIn profiles of Amazon employees. The results will be obviously different compared with the results of a search for the keyword *Amazon*. The results are going to match the query precisely. (Note that the LinkedIn site has become quite inconsistent over the last year. Only some LinkedIn profiles have the structure.)

As another example, GitHub organization pages include *Organization* (object) with firleds *<Programming Language>*, <Location>, and <URL>. (carrying values). Accordingly, you can set a CSE to search for organizations in Germany writing in Python.

It is impossible to put together a tip sheet with all CSE advanced operators because they are specific to each site and type of page. (For example, you can search for *Code* on GitHub but not on LinkedIn.) Operators depend on the information within the source code. A page may contain *Person, Code,* or both and even several objects of the same kind. An included object will have some fields with values such as location, job title, or employer for the Person object.

Typically, pages of the same kind on the same site, such as public profiles on a social site, have the same structure. They contain the same objects and values filled out. (As you know, there are exceptions.) The ability to query a site's structure through a CSE provides you with a filtered search of the site.

For any "structure-friendly" site with public profiles, using the special operators would achieve some of the following advantages:

- Additional – structure-dependent – search filters.

- Avoiding logins and going to the site (useful for LinkedIn since its profile search and viewing ability are limited, depending on the account).

- Better results visibility (for example, you can find LinkedIn profiles of people out of your network).

- Straightforward results collection (including scraping and processing in Excel).

The CSE search operators are not for the faint of heart: they look complicated, and you have the additional task of figuring out their exact syntax for each site. But the benefits are incredible. We will tell you all about finding hidden structured data of any web page and operators to query it.

THE PERSON OBJECT: A CLOSER LOOK

Since we often search for professionals, the *Person* object is of most interest to us.

Schema.org defines objects like *Person* as a list of fields those objects may potentially contain. Some fields contain data of a certain type, such as a string, number, or date. Other fields contain Schema.org objects (such as *Organization*, *PostalAddress*, or *Occupation*), which in turn contain other fields or objects (Figure 14.1):

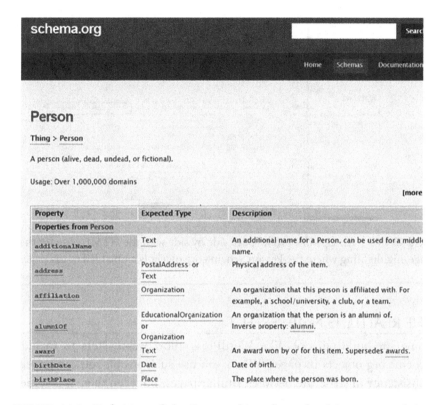

FIGURE 14.1 Definition of the Person object from the Schema.org website, showing some of the data fields a Person object can contain. Are you wondering what "undead" means? Schema.org definitions sometimes include references to nerdy terminology with humorous intent.

Below is a screenshot of a page with a Person object (Figure 14.2). To the right is the HTML code from that page defining the Person object. This code informs search engines of the object's presence and values. Don't panic! We share the screenshot for illustration purposes only: you do not need to know HTML to search using Schema objects.

FIGURE 14.2 An Upwork profile page side by side with the HTML code for that page, highlighting where the Person Schema.org object is embedded in the page.

THE REALITY IS NOT PERFECT

Some site-building tools like WordPress allow webmasters to include Schema.org objects in pages. When you use such a tool, you can expect consistency in structure between similar pages. But even then, it is sometimes not the case for unknown reasons.

While it would be nice if each web page containing a Schema.org Person object had all 68 fields defined in https://schema.org/Person, that is seldom the case. Schema.org objects embedded in pages usually contain a smaller subset of the possible values. Implementation varies from site to site. For example, a Person object will typically – though not always – include the person's name, company, title, and location but very rarely, an email address or gender. To determine which values are present in the Shema.org objects for a site, you need to study real (or cached) pages, not just the theoretical object definition, and use tools.

In addition to variations in the same object implementations between sites, there are often deviations between individual pages within the same site. Sometimes, it is because of a later site redesign. In this case, Google discovers the updated structure when it visits updated pages, so it takes time, and you need to use different operators to find pages with the same values.

But sometimes, pages are just inconsistent. For example, some LinkedIn profiles have no Person object, while others that have it are missing *person-org* (company or school), *person-role* (headline), or *person-jobtitle* (job title) values. (We thought this was due to LinkedIn public profiles redesign, but Googlebot still picks some updated profiles that communicate structure to it.)

Why some sites are inconsistent is hard to explain. Additional web pages' structure "oddities" include the following:

- Duplicating the same value in different fields (for example, each profile would have equal values in person-role and hCard-title)

- Using fields for purposes other than for which they were designed (for example, the field for the organization would store the person's address).

It is a bit wild out there! Start researching, and you will see what we mean. Here is an example screenshot of value redundancy within a structure (Figure 14.3):

```
"hcard": [
  {
    "fn": "▀▀▀▀▀▀▀▀▀",
    "photo": "https://www.xing.com/image/c_e_5_9b42d18f6_34312251_2/▀▀▀▀▀▀▀-foto.256x256.jpg",
    "title": "SAP CRM Senior Consultant"
  }
],
"person": [
  {
    "role": "SAP CRM Senior Consultant",
    "org": "Software",
    "location": "Winterthur"
  }
],
```

FIGURE 14.3   JSON code returned by Google's CSE API showing data for search results, demonstrating that some website owners implement metadata in unexpected ways.

NON-STANDARD OPERATORS

Even that is not the whole story! Some sites have pages with fields that are not even defined in Schema.org (for example, within the Person object). CSEs can search for those too – with "extremely" unique operators that work for those specific sites only.

Here are examples of *more:p:* operators that bring up the Person object with fields that are not in Schema.org's definition (we are using the Search Everything CSE):

- more:p:person-age

- more:p:person-nickname

- more:p:person-dollar OR more:p:person-euro

- more:p:person-linkedin

- more:p:person-me

- more:p:person-paypal

- more:p:person-account

You will need to look at the results and examine the Person fields to see what the fields mean (i.e., what type of information they contain and in what format) on each site.

HOW TO IDENTIFY STRUCTURED DATA ON A PAGE

The structured data we are talking about is invisible when looking at a page in a browser. You need to use tools to examine it. There are several tool options:

1. Use Google Structured Data Testing Tool (Paste a Page URL)
Many other structured data testing tools are available. Note that different tools show different info for the same page. The differences are due to the variety of ways pages "express" the structure in the source code. Some structure-testing tools "understand" different objects and values than others.

2. View Page Structure Using CSEs
You can set up a CSE so that it displays links to the results' structure (Figure 14.4):

Programmable Search

New search engine

Edit search engine

Linkedin

Setup
Look and feel
Search features
Statistics and Logs

Help
Visit Help Forum
(Ask a question)
Send Feedback

Promotions   Refinements   Autocomplete   Synonyms   **Advanced**

Enable or disable advanced search features.

› Results sorting ⊙   ON

↓ Websearch Settings ⊙

Refinement Style ⊙   ⦿ Links   ○ TAB

Results Browsing History ⊙   ○ Enable ⦿ Disable

Structured Data in Results ⊙   ⦿ Enable ○ Disable

No Results String ⊙

FIGURE 14.4 The CSE Control Panel showing the Advanced tab of the Search features page, highlighting the setting option to include a link to view structured data with each search result.

With the setting, each result will be accompanied by a link to view the page structure. For example, this search result has a *Person* object with the fields *role* and *org* and an *hCard* object with the fields *fn* and *title* (Figure 14.5):

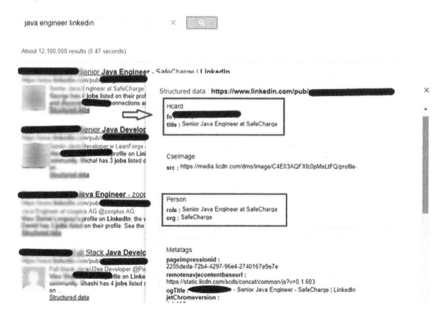

FIGURE 14.5 Sample search results for a CSE with the setting shown in Figure 14.4 enabled, highlighting the structured data display.

3. Structure-viewing tools and previews using CSEs, however, sometimes provide inaccurate or incomplete information regarding the
CSE operators' syntax. You can still use the tools, observe the structure, guess the CSE operators' format, and test whether they work. In
many cases, they would. But the best way to figure out exactly how to
write operators for a given site is to use the CSE APIs. APIs provide
complete information. We will tell you about the APIs later in the
book.

HOW TO SEARCH FOR STRUCTURED DATA WITH ANY CSE – ADVANCED SYNTAX

Special CSE operators depend on the website and structure of its pages.
More specifically, operators depend on Schema.org, Microformats,
Metatags, and other objects and values included in the pages' source code.

The operator formats are as follows:

```
more:pagemap:<data-type-name>
```

where data-field-name is an object like Person – finds pages containing the
object.

```
more:pagemap:<data-typefield-name>-:<data-field-
namevalue>
```

where data-field-name is an object like Person – finds pages containing the
object in which *<data-value>* is filled out (for example, a page has a *Person*
object with the field "gender" filled out).

```
more:pagemap:<data-type-name>-<data-field-
name>:<value>
```

where *<data-field-name>* is an object like *Person*, *<data-value>* is a value,
such as "org" (i.e., organization, a Person's employer), and the value is a
string like *IBM*. A complex operator such as this can find pages with the
object Person for whom the "org" value contains "IBM."

An alternative syntax that we will use going forward has just *p* instead of *pagemap* and works the same:

```
more:p:<data-typefield-name>-<data-field-
namevalue>:<value>.
```

Google.com doesn't "understand" the *more:p:* search syntax, but any Google Custom Search Engine does.

If you think these operators are new additions to CSEs, you are mistaken. Google mentions them <u>in 2009</u> and <u>2012</u>. Many people have been missing this capability, and it is time to learn it!

MORE: OPERATORS FOR REFINEMENTS

Note that the end-user can also write (different) *more:* operators for Refinements instead of selecting refinements from the UI. You can use the syntax *more:<refinement name>* for that purpose.

Here is an example using the Documents by Format CSE (defined in Chapter 6):

"member list" association healthcare <u>more:excel</u>.

While the previous query works the same as selecting the Excel refinement tab, this syntax allows you to search for a combination of refinements, for instance:

"member list" association healthcare <u>more:excel OR more:PDF</u>.

HOW TO FIND OBJECTS

According to the syntax rules, *more:p:person* added to a CSE search will narrow results down to pages with a Person object. For example, this query in the "Search Everything" CSE – <u>site:linkedin.com more:p:person</u> – will search for LinkedIn pages with a Person object. It works the same as narrowing to the Person object when creating a CSE.

Compare the results you get with and without narrowing down to a Person – <u>site:linkedin.com more:p:person</u> vs. <u>site:linkedin.com</u>. In the first case (left on the screenshot), the end-user is only getting profiles and not other linkedin.com pages. On the right, the results include pages without the *Person* object (Figure 14.6).

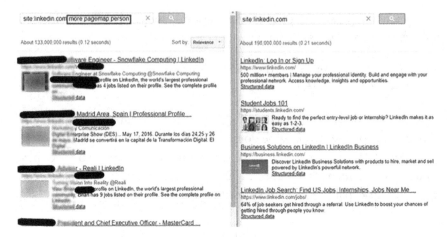

FIGURE 14.6 Demonstration of the *more:pagemap:* search syntax. On the left, sample search results for pages from LinkedIn that contain the *Person* Schema.org object, which are all user profiles. On the right, the same search without the *more:pagemap* operator finds results that are not profiles.

TIP: SEARCH FOR PROFILES ON A SOCIAL SITE

The Control Panel allows setting the type of object to which the CSE will narrow search results. As in the cases of LinkedIn and GitHub, searching for the Person object while X-Raying any social site usually narrows results to profiles. The mechanism gives you a quick way to set up a profile search for many sites.

Some sites that include the *Person* on public profiles, besides LinkedIn, are the following:

- Slideshare
- XING
- GitHub
- ResearchGate
- HackerRank
- Google Scholar
- ZoomInfo

- Vitals

- Doximity.com

(Note that neither Twitter nor Facebook includes objects in public profiles.)

FIND WHICH SITES AND PAGES CONTAIN WHICH OBJECTS WITH "SEARCH EVERYTHING"

The "Search Everything" CSE is your tool to find which sites have which structure. You can use *more:* operators on the "Everything" CSE to discover pages the *Person* or other types of objects on the sites you wish to search. To do so, use the *site:* operator for the target site together with exploratory *more:* statements looking for Objects. The previous example, site:linkedin.com more:p:person, shows how to find pages containing *Person* on LinkedIn.com. Here are some other examples (Figure 14.7):

- site:doximity.com more:p:person

- site:vitals.com more:p:physician

- site:linkedin.com more:p:jobposting

FIGURE 14.7 Another demonstration of the *more:pagemap:* search syntax. From left to right, showing sample search results containing the Person Schema. org object from Doximity, results containing the Physician Schema.org object from Vitals, and results containing the JobPosting Schema.org object from LinkedIn.

For the same site, narrowing to different objects will, of course, produce different search results for the same search string. Here are some examples to illustrate:

- LinkedIn/Person CSE search: <u>vp manufacturing</u>
- LinkedIn/JobPosting CSE search: <u>vp manufacturing</u>
- LinkedIn/Person CSE search: <u>3d printing</u>
- LinkedIn/BlogPost CSE search: <u>3d printing</u>

You can also search for Objects with a *more:* operator *without* narrowing to a site using "Search Everything." Not restricting to any site is a way to discover sites with a particular structure of interest. For example, you can search for all pages with a value present in the Person object's email address field with <u>more:p:person-email</u>.

NEXT: SEARCH BY FIELDS AND VALUES

You are ready to explore the most exciting part – searching for objects' values. As a reminder, advanced CSE operators have this format: *more:p:<data-field-name>:<data-value>*.

Here is an example of a search looking for an object value using the Search Everything CSE: <u>site:linkedin.com more:p:person-org:IBM</u> (Figure 14.8).

FIGURE 14.8 A demonstration of searching for data field values within the Person Schema.org object.

This search will look for pages containing a *Person* object whose *org* (short for organization) field value includes the word *IBM*.

MULTIPLE OBJECTS AND INSTANCES IN ONE PAGE

One unfortunate challenge we face with CSE operators is that they would look for values for any object of a specified kind with a given field value in a page. There could be two different objects on one page you are querying or two values of the same kind within the same object. In either case, expect to find false positives.

For example, if you search for *more:p:person-name:mary*smith*, you will also find pages with two Person objects – for *Mary Kennedy* and *Harold Smith*. The same is true about other values, such as addresses, job titles.

Indeed's job posts contain multiple *jobposting* instances and multiple values for posting organizations for each instance, as a JSON output shows (Figure 14.9):

```
"jobposting": [
    {
        "hiringorganization": "DuPont",
        "name": "DuPont",
        "description": "The PRE Leader also provides process and product technology development and operations support
services, focused on meeting customer needs and business...",
        "title": "Production Engineer - Polymers"
    },
    {
        "hiringorganization": "Honeywell",
        "name": "Honeywell",
        "description": "Bachelor's Degree in Chemical or Mechanical Engineering. We have an exciting opportunity
available for a Production Engineer located in Colonial Heights, VA....",
        "title": "Production Engineer"
    },
    {
        "hiringorganization": "AdvanSix",
        "name": "AdvanSix",
        "description": "Manage the execution of small capital investments as project engineer through their entire life
cycles in areas assigned:....",
        "title": ".Project Engineer"
    },
    {
        "hiringorganization": "AdvanSix",
        "name": "AdvanSix"
```

FIGURE 14.9 JSON code returned from Google's CSE API, highlighting the JobPosting Schema.org data in a job posting from Indeed.

There is nothing you can do about the false positives. You will need to review results and filter out the "wrong" results.

Luckily for us, the majority of the pages of interest include only one Person object. The web is not as accommodating for job posts: job post aggregators like Indeed have lists of jobs on public pages containing multiple JobPost objects.

You can combine terms (search for each of them) within the same operator by using an asterisk (*) between the words. You can search for one or more terms by separating them with a comma (,).

Examples follow.

ORDER OF BOOLEAN OPERATIONS

As stated earlier, the comma (,) stands for the Boolean "OR", and the asterisk (*) means "AND." (There is no way to search for a phrase.)

Examples:

- more:p:person-location:california,oregon

- more:p:person-org:cisco,workday

- more:p:person-name:henry*jones

- more:p:person-title:social*media*marketing

Very interestingly, *more:* operators support a different order of Boolean operators compared to Google. On Google, ORs have the highest priority. On CSEs, ANDs have the highest priority. Consider an example: more:p:person-jobtitle:microsoft*developer,google*lead,amazon*manager. It is a search for people with the job title containing the following:

- (microsoft AND developer) OR

- (google AND lead) OR

- (amazon AND manager)

CSE OPERATORS FOR LESS TECHNICAL END-USERS

While less technically savvy people will likely refuse to write advanced CSE operators, you can serve these users in one of the two ways:

1. Create CSEs that automatically add CSE operator-based expressions to the search.

Unfortunately, there is no straightforward way to insert the user's input as parameters for operators. But you can narrow down to types of objects (e.g., *Organization*) and values (strings such as *IBM*) as a CSE auto-add string setting. Here is an example of such a CSE.

The Search for Females CSE automatically adds *more:p:metatags-profile_gender:female* to every search. Not every site uses the *profile_gender* metatag, but here are three popular sites that do:

1. ResearchGate (scientific papers)

2. Doximity (a healthcare site)

3. SpeakerHub (speakers)

User profiles on each of the three sites contain the *profile_gender* metatag. Due to the auto-appended search operator, only women's profiles will be included in the results of the following example searches (Figure 14.10).

Example searches:

- <u>non-profit board member</u>

- <u>sorority founder</u>

- telecommunications

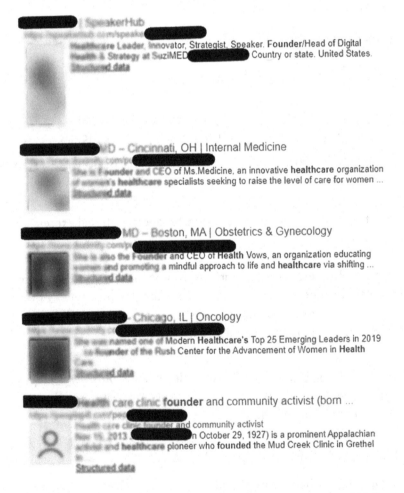

FIGURE 14.10   Sample search results from the Search for Females CSE, demonstrating that all the results are, in fact, women.

2. Use CSE APIs to run your software code, hide operators from the end-user, and fetch values from the structure.

Our subscription-based tool, <u>Social List</u>, does that. It runs *more:* searches combined with user-entered values. (You cannot achieve that with CSEs.) Social List makes advanced operators invisible to the user who only enters values into the fields and sees results that exactly match the input. For instance, the user can find XING.com members who live in a given city such as Berlin or Google Scholar profiles with a verified email at *berkeley.edu.*

CSE OPERATORS FOR THE OPEN WEB

To discover sites that provide specific structure and field values, you can search with operators based on Schema.org objects without narrowing to a site and reviewing the sites that come up. In many of our research projects, we did not know which sites have "our" results ahead of time and got to learn about them in the search results. The searches we ran included our target keywords and CSE operators for the *Person, Physician, Dentist,* or other object types. We then investigated the most commonly appearing sites and learned about the full structure they provide. For many data-rich sites, we created dedicated CSEs.

The following are examples of searching for an object and value "on the open web," not narrowing to a site:

- <u>more:p:person-knowsabout:search</u>

- <u>more:p:person-nationality:russia</u>

- <u>more:p:person-taxID</u>

- <u>more:p:person-weight:120</u>

- <u>more:p:person-alumniof:harvard</u>

- <u>more:p:person-birthplace:madagascar</u>

- <u>more:p:person-description:veteran</u>

- <u>more:p:physician-telephone:303</u>

- <u>more:p:physician-hospitalaffiliation:mount*sinai</u>

- <u>more:p:physician-medicalspecialty:oncology</u>

- more:p:physician-address:houston

- more:p:physician-aggregaterating:5

- more:p:physician-isacceptingnewpatients:true

- more:p:physician-department:urology

- more:p:physician-email:org

- more:p:physician-description:surgeon

- more:p:physician-member:society,association

- more:p:physician-taxID

- more:p:physician-pricerange:150,200

- more:p:physician-address:chicago

- more:p:physician-department:medical

- more:p:physician-numberofemployees:3,4,5,6

You can use objects' definitions on Schema.org, guess different common field values, and run combinations of operators and keywords to bring up various sites. You can extend the list of sites by excluding the sites already on your list and re-running the queries.

These niche sites may be hard to discover by simple searches. As an example, run one of the previous examples – more:p:person-alumniof:harvard. The sites that show up in the results – *mylife.com, dealroom.com, helloenglish.com, wikilawschool.net, rusprofile.ru* – are new to us. Each of them has valuable data and structure to query. Some deserve a dedicated CSE, depending on the research goals.

CHAPTER 15

API and Other Considerations

CUSTOM SEARCH ENGINE APIs

The <u>API</u> (which stands for Application Program Interface) queries CSEs from software code or browser address bar. If you are not a coder, you can run API calls from your browser's address bar, one at a time, or use Google's Structured Data Testing Tool.

Using the APIs requires obtaining a key (a long unique string of symbols) from Google for your subsequent use. Generate your key from <u>this page</u> by scrolling down to find and press the "Get a Key" button (Figures 15.1 and 15.2).

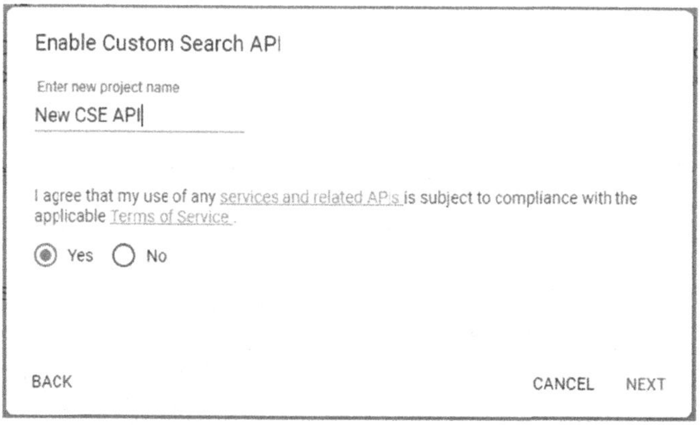

Enable Custom Search API

Enter new project name
New CSE API|

I agree that my use of any <u>services and related APIs</u> is subject to compliance with the applicable <u>Terms of Service</u>.
⦿ Yes ◯ No

BACK   CANCEL   NEXT

FIGURE 15.1   Step one to enable the CSE API for your account.

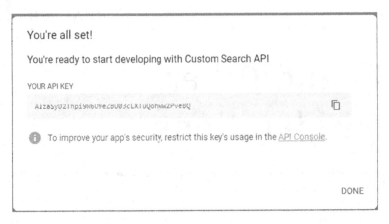

FIGURE 15.2   Continuing from Figure 15.1, step two of enabling the CSE API on your account – getting your API key.

Input for an API call is a key, a CSE ID (a value you can copy from the Control Panel), a query string, and (optional) parameters. The output comes in JSON format, which is easy to parse.

You can run an API query from your browser in the following fashion: https://www.googleapis.com/customsearch/v1?key=KEY&cx=CSE ID&q=QUERY.

The URL will look similar to this (Figure 15.3):

https://www.googleapis.com/customsearch/v1?key=AIzaSyA_Ee1WVja
HltF6B6wr2mkONXetYsy-
ogU&cx=01679963297930850473:buxbhwd9nxm&q=a

FIGURE 15.3   A complete Google CSE API call, including the API key, CSE ID, and query string.

You can also run a query from here: https://developers.google.com/custom-search/v1/reference/rest/v1/cse/list. The interactive dialog on the left is the easiest way to run an API call since you only need to enter parameters (Figure 15.4):

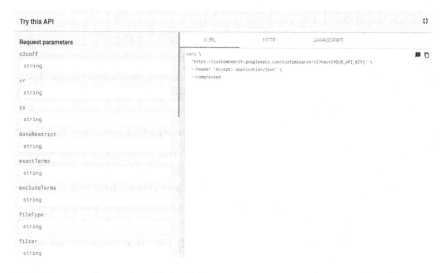

FIGURE 15.4 Screenshot of the CSE API testing screen from Google's support documentation.

The UI in the dialog may look unfamiliar, but it is as user-friendly as you can get. Try it out. It is your gate to the operator syntax investigation.

An API call produces a JSON-formatted output page you can analyze to figure out the correct operator formats.

Tip: The APIs provide the most accurate source of info on writing CSE operators compared to Google Structured Tool, other tools, and structure previews in CSE search results. This was one of our most significant discoveries after extensive tests of structure-viewing tools, none of which fit the bill. (The very first discovery was that special operators work on *any* CSEs.)

So, even if you do not intend to write code, running API calls in a browser or via Google's tool is necessary when you are figuring out the right *more:* operator syntax for a site.

STEPS TO IDENTIFY ADVANCED *MORE:P:* SYNTAX

The exact syntax of the CSE operators depends on the objects of a site's pages and their values. Since not every page includes all values that may be present for a particular site, it is a good idea to review several pages' structures to identify a complete set of operators. Run diverse searches and review varied pages.

If you look at the structured data preview with a CSE, you may be able to figure out how to write operators. For example, you can expect *more:p:person-name*, *more:p:person-jobtitle*, and *more:p:person-org* to work.

However, a CSE with this setting, "show structure," may not show the full structure to query. You must run several API calls, and review the results of each, to find out the exact operator formats. Here is a sample fragment of a JSON output from an API call. The structure you can query is the section "pagemap" (Figure 15.5):

```
  "htmlFormattedUrl": "https://www.xing.com/profile/Stefan_Nowak3",
  "pagemap": {
    "hcard": [
      {
        "fn": "Stefan Nowak",
        "photo": "https://profile-images.xing.com/images/4abe39ce122db218f5e244adad573d6d-2/stefan-
nowak.256x256.jpg",
        "title": "Managing Delivery Architect | People Development Lead"
      }
    ],
    "cse_thumbnail": [
      {
        "src": "https://encrypted-tbn1.gstatic.com/images?q=tbn:ANd9GcRgegCgcliY732RZAu7mQuW-
3SqGrwTkosgHRZZGDFMQSt1ZoM6Z60wGt8",
        "width": "225",
        "height": "225"
      }
    ],
    "person": [
      {
        "role": "Managing Delivery Architect | People Development Lead",
        "org": "Internet und Informationstechnologie",
        "location": "Berlin"
      }
    ],
    "metatags": [
      {
        "logjam-action": "Frontend::ProfileLo::ProfilesController#show",
        "og:image": "https://profile-images.xing.com/images/4abe39ce122db218f5e244adad573d6d-2/stefan-
nowak.1024x1024.jpg",
        "twitter:card": "summary",
        "twitter:title": "Stefan Nowak - Managing Delivery Architect | People Development Lead - Capgemini",
        "og:type": "profile",
        "logjam-request-id": "profilesloggedout-production-ad45a19379b94474bfc6be2c82ddd857",
        "og:site_name": "XING",
        "og:title": "Stefan Nowak - Managing Delivery Architect | People Development Lead - Capgemini",
        "csrf-param": "authenticity_token",
        "og:description": "Managing Delivery Architect | People Development Lead",
        "twitter:image": "https://profile-images.xing.com/images/4abe39ce122db218f5e244adad573d6d-2/stefan-
nowak.1024x1024.jpg",
        "referrer": "default",
        "twitter:site": "@XING_de",
        "viewport": "width=device-width, minimum-scale=1, maximum-scale=1, user-scalable=0",
        "twitter:description": "Managing Delivery Architect | People Development Lead",
```

FIGURE 15.5   JSON code returned from Google's CSE API, highlighting the segments representing data that can be searched with the *more:pagemap:* operator.

By looking at this output, you can derive the syntax of the CSE operators (Figure 15.6). For example, this piece of a JSON output –

```
"pagemap": {
  "cse_thumbnail": [
    {
      "src": "https://encrypted-tbn1.gstatic.com/images?
ɔn:ANd9GcTOQos4IjTqTxIAms29_uCb7jddm7gSuleITkfkrnrU5m5cwjDYLOx2u70",
      "width": "225",
      "height": "225"
    }
  ],
  "breadcrumb": [
    {
      "title": "Doximity",
      "url": "Doximity"
    },
    {
      "title": "States",
      "url": "States"
    },
    {
      "title": "Texas",
      "url": "Texas"
    },
    {
      "title": "Dallas",
      "url": "Dallas"
    }
  ],
  "person": [
    {
      "image": "https://doximity-res.cloudinary.com/image/upload/t_
:eholder-unregistered-1.jpg",
      "jobtitle": "Julia Gregg is a psychiatric-mental health nurse
iliated with Medical City Dallas and Medical City Green Oaks Hospital
      "name": "JuliaNicoleGreggNP"
    }
  ],
  "postaladdress": [
    {
      "addresslocality": "Dallas",
      "postalcode": "75230",
      "addressregion": "TX",
      "addresscountry": "United States",
```

FIGURE 15.6   A zoomed-in look at the top part of the code shown in Figure 15.4
highlighting the Person and PostalAddress Schema.org object data.

– tells you that the end-user can write the operators:

- *more:p:person-org*
- *more:p:person-name*
- *more:p:person-jobtitle*
- *more:p:postaladdress-addresslocality*
- *more:p:postaladdress-postalcode*
- *more:p:postaladdress-addressregion*
- *more:p:postaladdress-addresscountry*

This piece, within another result, (Figure 15.7) –

```
"hcard": [
    {
    "fn": "Rosalee Parkins,"
    "title": "Social Worker,"
    "photo": "Rosalee Parkins placeholder image"
    }
```

FIGURE 15.7 Another segment of the code from Figure 15.4, highlighting the hCard microformat data.

– tells you that the end-user can apply:

- *more:p:hcard-fn*
- *more:p:hcard-title*
- *more:p:hcard-photo*

For each site you investigate, we recommend running various keyword queries and reviewing the JSON results to verify your operator syntax guesses' correctness and completeness. Then search the site using relevant CSE operators and check whether results match your required values.

META TAGS

In LinkedIn public profiles, the Meta tags section does not have anything unique to query. It is different for GitHub. Meta tags on GitHub profiles contain a bio string, highlighted in this screenshot (Figure 15.8):

FIGURE 15.8 Screenshot of a GitHub user profile highlighting the bio string, which cannot normally be targeted explicitly in searches.

The "Meta tags" part of JSON output for the profile (Figure 15.9):

```
"metatags": [
  {
    "og:image": "https://avatars1.githubusercontent.com/u/447033?
u=f58e3859c7c2fe52b38a94efbe2b9144e1e38995&v=4",
    "twitter:card": "summary",
    "theme-color": "#1e2327",
    "og:site_name": "GitHub",
    "octolytics-event-url": "https://collector.githubapp.com/github-external/browser_event",
    "optimizely-sdk-key": "cowimJNste4j7QnBNCjaw",
    "html-safe-nonce": "6791d5c4a27347e7ee9e5afcd3563ca8af04b109",
    "expected-hostname": "github.com",
    "octolytics-app-id": "github",
    "og:description": "Works at LinkedIn. Lives in Berkeley. Likes a nice hike. - garris",
    "browser-errors-url": "https://api.github.com/_private/browser/errors",
    "hostname": "github.com",
    "twitter:site": "@github",
    "octolytics-dimension-ga_id": "1145589619.1589760000",
    "google-analytics": "UA-3769691-2",
    "browser-stats-url": "https://api.github.com/_private/browser/stats",
    "visitor-payload":
"WZlcnJlciI6IiIsInJlcXVlc3RfaWQiOiJBNjU1OjA1NEQ6MUQxRDY6Mzg1MTA6NUVDMkQwNEEiLCJ2aXNpdG9yX2lkIjoiMTI
4QOTA1Nzg4Ni9IsInJlZ2lvbl9lZGdlIjoiaWFkIiwicmVnaW9uX3JlbmRlciI6ImlhZCJ9",
    "enabled-features": "MARKETPLACE_PENDING_INSTALLATIONS",
    "twitter:title": "garris - Overview",
    "og:type": "profile",
    "profile:username": "garris",
    "og:title": "garris - Overview",
    "visitor-hmac": "f2cc25aa5d5161ac6d9e96bdf97fd96a3cd7b582f277c1cb4ec08408a9e2fe9d",
    "request-id": "A655:054D:1D1D6:38510:5EC2D04A",
    "analytics-location": "/\u003cuser-name\u003e",
    "twitter:image:src": "https://avatars1.githubusercontent.com/u/447033?
u=f58e3859c7c2fe52b38a94efbe2b9144e1e38995&v=4",
    "fb:app_id": "1401488693436528",
    "viewport": "width=device width",
    "twitter:description": "Works at LinkedIn. Lives in Berkeley. Likes a nice hike. - garris",
    "dimension1": "Logged Out",
    "octolytics-host": "collector.githubapp.com",
    "og:url": "https://github.com/garris",
    "dimension3": "mobile"
  }
]
```

FIGURE 15.9   JSON code returned from Google's CSE API, highlighting the Meta tags data for the profile shown in Figure 15.7.

Meta tags repeat the GitHub bio value twice, while it is nowhere else in the JSON output.

You can write CSE operators to query values stored in Meta tags, as you can for fields of Schema.org objects. Here are some example searches:

- more:p:metatags-og_description:clojure,lisp

- more:p:metatags-og_description:love*python

- more:p:metatags-og_description:azure*devops

- more:p:metatags-og_description:work*at*facebook

- more:p:metatags-og_description:caltech*graduate

- more:p:metatags-og_description:open*to*new

Notice how *og:description* in JSON is converted to *og_description* in the operator (Figure 15.10):

FIGURE 15.10 CSE search results demonstrating the ability to target the bio string in GitHub user profiles with the *more:pagemap:* operator.

SORTING RESULTS

After performing a search via the Public URL for a CSE, including apply-ing a refinement, you can sort the results. By default, CSEs sort results by relevance, just like Google. The other default option included in a CSE is to sort results by date. You can add custom sorting options, or remove default options, via the Control Panel under Search features -> Advanced - > Results sorting (Figure 15.11):

Promotions	Refinements	Autocomplete	Synonyms	**Advanced**

Enable or disable advanced search features.

▾ **Results sorting** ❷   **ON**

Add key	Delete

☐ Sort by ... ❷	Label ❷
☐	Relevance •
☐ date	Date

FIGURE 15.11 The CSE Control Panel, showing the Advanced tab of the Search fea-tures page, highlighting settings to change search result sorting behavior for a CSE.

When adding a custom sorting option, you can include any structured field that would work with the *more:pagemap:* operator, so long as the values in that field are numeric. So, for example, if a page includes the Person Schema.org object with the data field *telephone*, you can search for the field in a CSE using *more:p:person-telephone*. You would enter person-telephone as the key to sort by and an appropriate label like *Phone*. When using the CSE from the Public URL, you would see a new *Phone* option under the same sort of drop-down as *Relevance* and *Date*. Choosing this option will cause the results to be sorted (in descending order, so *609-555-1212* will come before *510-555-1212*) by the telephone numbers. Results that do not include a *person-telephone* data field will be excluded in that case (just as all pages without Date info are excluded if you choose the *Date* default sorting option). Sometimes you will even see *No results* after you change the sort-ing option. That means that none of the search results contained the rele-vant metadata. The requirement for the sorting fields to be numeric is strict – any non-number, non-punctuation value in the field (even, for example, as in a time written as 7 pm) will make sorting by that field impossible.

4

How We Built Our CSEs

You will not learn to create CSEs only by understanding the theory, definitions, and syntax rules. You need to get hands-on experience building and using CSEs to truly "get" them. It is just like swimming, dancing, or cooking.

Learning CSE creation needs experiences, not unlike <u>kinesthetic learning</u>. We think the same applies to learning how to do any research on the Internet.

For each CSE below, we explain how we constructed it so that you can follow our steps, and provide a usage screenshot and sample search links. The underlined titles point to the CSEs' Public URLs.

We encourage you to reproduce these CSEs and create your own variations!

CHAPTER 16

Basic CSE Examples

GOOGLE SCHOLAR PROFILES CSE

Google Scholar is a freely accessible web search engine that indexes the full text or metadata of scholarly literature across various publishing formats and disciplines. Released in beta in November 2004, the Google Scholar index includes most peer-reviewed online academic journals and books, conference papers, theses and dissertations, preprints, abstracts, technical reports, and other scholarly literature, including court opinions and patents (*source:* Wikipedia).

The Google Scholar Profiles CSE includes the URL pattern *scholar. google.com/citations?user** in the list of sites to search.

Example searches:

- machine learning

- cancer clinical trails

- mass spectrometry

ABOUT.ME CSE

About.me is a personal web hosting service co-founded by Ryan Freitas, Tony Conrad, and Tim Young in October 2009. On February 21, 2019, About.me was acquired by Broadly.

The About.Me CSE includes the URL pattern *about.me/** in the lists of sites to search, and the URL pattern *about.me/*/** in the lists of sites to exclude.

Example searches:

- social media marketing

- chief happiness officer

- belgium business development manager

DIVERSITY ASSOCIATIONS CSE

The Diversity Associations CSE searches for results from .org websites, appending the keyword *association* to every search. It includes several *Refinement* tabs (see Figure 16.1). When you select one of these tabs, the CSE filters the results by applying an OR statement of relevant keywords. (Example – the refinement *Black* adds *Black OR African-American* to the search.)

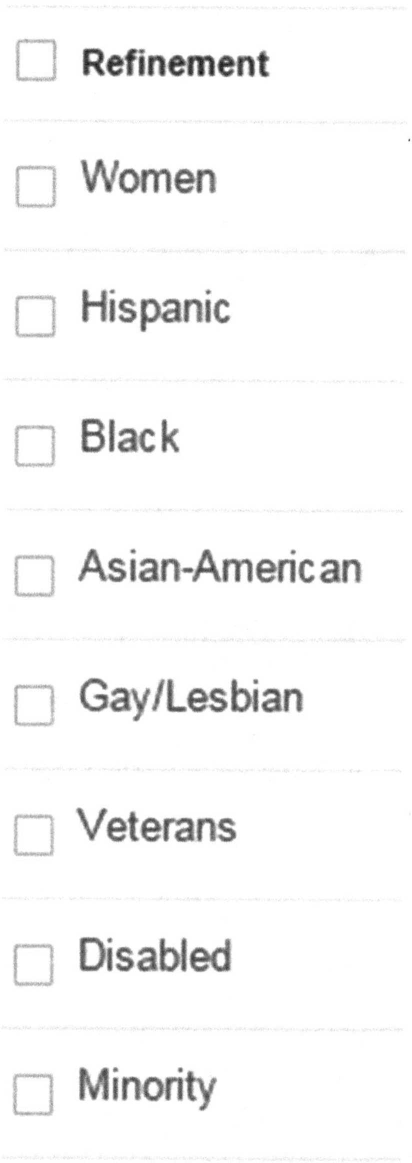

☐ **Refinement**

☐ Women

☐ Hispanic

☐ Black

☐ Asian-American

☐ Gay/Lesbian

☐ Veterans

☐ Disabled

☐ Minority

FIGURE 16.1 The CSE Control Panel, showing the Refinements tab of the Search features page, highlights the refinements of the Diversity Associations CSE.

Example searches:

- STEM

- technology

- accounting

- computer science

- MBA

- aviation

- customer support

- agency advertising

EMAIL FORMATS CSE

The Email Formats CSE searches public profiles of the B2B database ZoomInfo.com. ZoomInfo is an American subscription-based software as a service company based in Vancouver, Washington, that sells access to its database of information about business people and companies to sales, marketing, and recruiting professionals. CEO Yonatan Stern and Chief Scientist Michel Decary founded Zoom Information Inc. (ZoomInfo) in 2000 as Eliyon Technologies (*source:* Wikipedia).

The CSE includes *zoominfo.com/p.* (Public ZoomInfo profiles "live" under this domain.)

Auto-appends the key phrase *"email * * com"* to every search (anticipating finding phrases containing .com email addresses).

To use, search for a corporate email domain in the quotation marks (most often being the same as company domain, e.g., "ge.com") to see some representative unmasked company email addresses and potentially figure out the corporate email format.

Example searches (Figure 16.2):

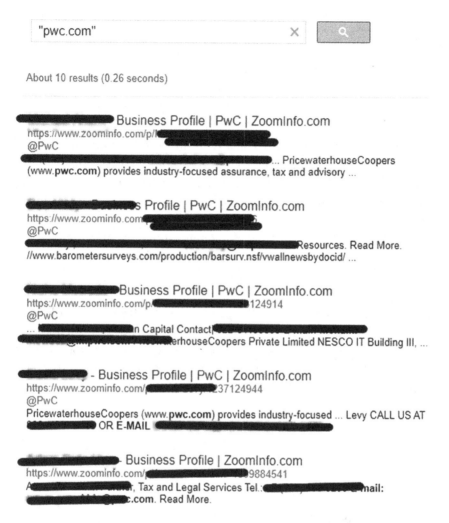

FIGURE 16.2 Sample search results for the Email Formats CSE, demonstrating how reviewing multiple email addresses from the same company can quickly reveal the company's format.

- "ge.com"

- "hilton.com"

- "stryker.com"

- "pwc.com"

You might be wondering why you are finding visible email addresses on a subscription-based site that provides contact info for pay. The reason is that some ZoomInfo profile pages include cached quotes of news relevant to the person. The cached content sometimes contains contact information.

Note that Google may have "forgotten" about the past news while ZoomInfo keeps the history, which could be useful for your research.

HOOVERS CSE OPERATORS

D&B Hoovers was founded by Gary Hoover and Patrick Spain in 1990 as an American business research company that provides information on companies and industries through their primary product platform named "Hoover's." In 2003, it was acquired by Dun & Bradstreet and operated for a time as a wholly owned subsidiary (*source:* Wikipedia).

Use the Hoovers CSE.

Example searches:

- manufacturing

- ernst and young competitors

- medical devices

LANGUAGE PROFICIENCY CSE

The Language Proficiency CSE finds results from LinkedIn.com and adds the phrase "Native or bilingual proficiency" to every search. Please keep in mind that a user may list more than one language at differing levels of proficiency. As a result, a search for *Bahasa Malay* using this CSE may return some profiles of users who list Bahasa Malay at Beginner proficiency and another language (such as English) at Native or bilingual proficiency.

Example searches:

- bahasa english

- georgian

- spanish french portuguese

- greek latin french dutch russian

- esperanto

- latin french spanish dutch

HIDDEN RESUMES CSE

The Hidden Resumes CSE has no included or excluded sites, and uses the "search the entire web" setting. The following query addition is appended to every search:

```
intitle:resume OR inurl:resume OR intitle:CV OR
inurl:CV -sample -example -samples -templates
-template -job -jobs -careers
```

Take a look at some examples on the site hiddenresumes.com.

DEVELOPER RESUMES CSE

The Developer Resumes CSE uses the "search the entire web" setting and includes results from GitHub.com. The following query addition is appended to every search: *github.com*; "search the entire web."

Query addition:

```
intitle:resume OR inurl:resume -job -jobs -hiring
-sample -example -samples -examples -inurl:issues
```

Example searches:

- C++ framework

- parsing algorithm

- PhD computer science
- PHP developer

URL SHORTENERS CSE

URL shortening is a technique on the World Wide Web in which a URL may be made substantially shorter and still direct to the required page. URL shortening services such as bit.ly provide the functionality.

The URL Shorteners CSE searches several popular URL shorteners, surfacing valuable links that someone had shared that have made it into Google's Index. The CSE can be useful to search for articles and publications that others have found worth sharing (Figure 16.3).

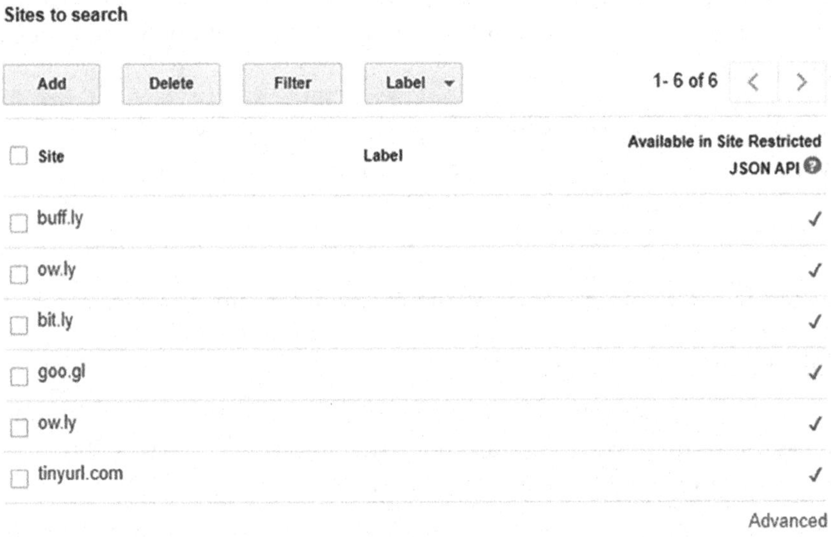

Sites to search

| Add | Delete | Filter | Label ▾ | | 1- 6 of 6 ‹ › |

Site	Label	Available in Site Restricted JSON API ❷
☐ buff.ly		✓
☐ ow.ly		✓
☐ bit.ly		✓
☐ goo.gl		✓
☐ ow.ly		✓
☐ tinyurl.com		✓

Advanced

FIGURE 16.3 The CSE Control Panel, showing the Basic tab of the Setup page with the Sites to search list for the URL Shorteners CSE.

Example searches:

- sourcing hacks
- SAT prep

- search engine marketing

- organic chemistry

SEARCH EVERYTHING CSE

The <u>Search Everything CSE</u> searches the entire web and has no restrictions.

You can use the CSE instead of Google. Its two advantages over Google are as follows:

1. CSEs show no "CAPTCHAs."

2. Additional search syntax (that we will discuss) is supported.

Here is how to create a CSE with no restrictions, like the "Search Everything" noted earlier. Create a CSE by specifying any site. In the Control Panel, choose "Search the entire web" and then remove the original included site from the CSE definition. You will have created a CSE that searches for "everything," as Google does.

CHAPTER 17

Object-Oriented CSEs

Y OU CAN ALSO ADD settings (other than including and excluding sites) to your CSE. This chapter's examples primarily include CSEs that rely on Schema.org objects being present in the results.

DEV.TO CSE

Dev.to (or just DEV) is a platform where software developers write articles, participate in discussions, and build their professional profiles. We value supportive and constructive dialogue in the pursuit of great code and career growth for all members. The site has 300K+ members (*source: site*).

The DEV CSE simplifies the process of finding user profiles on dev.to. Example search:

- #javascript amsterdam

There is not a whole lot of "structure" on the profiles (less than on GitHub's, for example), but you can search specifically for bios with a Custom Search Engine: more:p:person-description:java.

PHYSICIANS CSE

The Physicians CSE uses the "search the entire web" setting and features no included or excluded sites. Only pages that include a Schema.org *Physician* object will be shown in the results.
Example searches:

- neurosurgeon sydney

- pediatrician toronto

- phlebotomy chicago

- emergency medicine

ACCOUNTANTS CSE

The Accountants CSE uses the "search the entire web" setting and features no included or excluded sites. Only pages that include a Schema.org *Accountant* object will be shown in the results.
Example searches:

- CPA corporate tax denver

- tax audit kpmg

- outsource accounting "contact us OR me"

SOFTWARE CODE CSE

The Software Code CSE uses the "search the entire web" setting and features no included or excluded sites. Only pages that include a Schema.org *Code* object will be shown in the results.
Example searches:

- inurl:authors OR intitle:authors OR intitle:credentials

- django extensions

- intitle:resume OR intitle:CV

META TAGS

In addition to Schema.org, there are other standard formats that pages use to communicate structure and values to search engines. Some pages include structured data in only one format, and some pages include several instances of structure in various formats. It is all in the hands of webmasters.

You can search by fields that webmasters have included in Meta tags, for example:

- more:p:metatags-email
- more:p:metatags-phone
- more:p:metatags-tel
- more:p:metatags-age
- more:p:metatags-location
- more:p:metatags-linkedin
- more:p:metatags-mobile
- more:p:metatags-recruiter.

The list continues.

CHAPTER 18

Determining CSE Operators for Social Sites and *more:*

GitHub CSE OPERATORS

GitHub, Inc. is a US-based global company that provides hosting for software development version control using Git. In 2018, GitHub became a subsidiary of Microsoft for US$7.5 billion. It offers the distributed version control and source code management functionality of Git, plus its own features.

As of January 2020, GitHub reports having over 40 million users and more than 100 million repositories (including at least 28 million public repositories), making it the largest host of source code in the world (*source:* Wikipedia).

GitHub has internal advanced search operators such as *language:* and *location:* for user search and supports (nonstandard) Boolean search syntax. However, a GitHub search is limited, and you would complement it well by using a GitHub CSE.

We will use the Person Search - GitHub CSE. (It is a good idea to create your own GitHub CSE, try these strings and vary them.)

For searching GitHub, the operators are as follows:

- *more:p:person-homelocation* – A free-form text field into which the user may enter any value. Though most users enter a correct location, two people may enter the same location in different ways. For example, one person may write NYC, while another may write New York, NY.

- *more:p:person-name* – The name listed on the user's profile (may be a pseudonym, first name, full name, or whatever the user chooses).

- *more:p:hcard-url* – GitHub user profiles can include links to other sites, and you can search for these links with this operator.

- *more:p:person-worksfor* - member's employer (if listed)

Example searches:

- more:p:person-homelocation:amsterdam – living in Amsterdam

- django extensions – search by keywords

- intitle:resume OR intitle:CV – page has either *resume* or *CV* in the title

- more:p:hcard-url:linkedin – has a LinkedIn profile as the personal site URL

- more:p:person-worksfor:google more:p:person-homelocation:mountain* view – works for Google, lives in Mountain View

LinkedIn CSE OPERATORS

LinkedIn is an American business and employment-oriented online service that operates via websites and mobile apps. Launched on May 5, 2003, it is mainly used for professional networking, including employers posting jobs and job seekers posting their CVs. As of 2015, most of the company's revenue came from selling access to information about its members to recruiters and sales professionals. Since December 2016, it has been a wholly owned subsidiary of Microsoft. As of May 2020, LinkedIn had 690+ million registered members in 150 countries.

The company was founded in December 2002 by Reid Hoffman and founding team members from PayPal and Socialnet.com (*source:* Wikipedia).

LinkedIn is a site with the largest amount of self-entered professional data. It is a site we cannot ignore, despite its limitations, inconsistencies, user-unfriendly UI, bugs, expensive subscriptions, and dismal customer support.

X-Raying can nicely complement your LinkedIn searches since it will find more info (such as recommendations), show profiles in results regardless of whether they are in your network, and allow any number of searches free of charge.

Public LinkedIn profiles have the person's name, job title, and employer in the page title. With an X-Ray, you can search for job titles and companies using the operator *intitle:*. Examples:

- site:linkedin.com/in intitle:"chief marketing officer"

- site:linkedin.com/in intitle:"chief marketing officer"

However, for some terms (such as *SAP*, *Salesforce*, or *Oracle*), this method will be finding both company employees and people who have the term in their job title. So it will be getting false positives.

The CSE we will use is LinkedIn X-Ray (or you can run these queries in your own LinkedIn X-Ray CSE).

Very unfortunately, in the last six months or so, Google's and LinkedIn's software relationship around recognizing objects went sour. Some profiles still present a structure, but you will miss many matching results if you use operators that look for the *Person* object.

We do not know the reason for the inconsistency in the profiles' page structure. Google CSE support responds to questions about sites that requestors have built and is clueless in our case resolution since we did not build LinkedIn.com.

The operators (that only work for pages containing *Person*) are as follows:

- *more:p:person-role:<role>* (searches for profile headlines!)

- *more:p:person-org:<company>*

- *more:p:organization-name:<school>*

However, you can still search for headlines since they are present in Meta tags on all profiles! Example:

- more:p:metatags-og_title:hiring

The *more:p:person-role:* operator will also search in LinkedIn profile headlines (and not in job titles). LinkedIn does not provide this filter, even to its most high-paying subscribers. There's no way to accurately Google for headlines either. There is only one way to search for headlines – via Google CSEs.

Why is the headline search useful? The default LinkedIn headline has the format *<job title> at <company>*. But many members have customized the headline to say, *"open to new opportunities," "I am hiring,"* or *"good at Python."* Figure 18.1 shows a screenshot of a LinkedIn profile illustrating this point:

FIGURE 18.1 Partial screenshot of a LinkedIn user profile, highlighting the difference between a profile headline and the user's current job title.

Example searches:

- more:p:person-role:open*to*new – "open to new opportunities"

- more:p:person-role:hiring – "hiring"

- more:p:person-role:django*python – self-identified top skills example (looks for members who list skills in the headline and not necessarily a job title)

- more:p:organization-name:harvard – Harvard grads

- more:p:person-jobtitle:visionary – results are profiles where the headline contains *visionary*

- more:p:person-org:Swedbank more:p:person-org:developer,engineer – developers OR engineers working at Swedbank

- more:p:person-org:ibm more:p:person-role:partner (combining operators)

- more:p:person-org:morgan*stanley more:p:person-role:vice*president (combining operators)

SLIDESHARE CSE OPERATORS

As the name implies, Slideshare is a website for hosting and sharing PowerPoint and other presentation slides, along with other documents. Purchased by LinkedIn in 2012, then by Scribd in 2020, the site has about 70–80 million user accounts and hosts millions of documents. Documents uploaded to Slideshare are converted into images, used for viewing on the site. They have their text contents extracted and presented inline below the images of slides or pages from the document. Due to this conversion, an effective strategy for finding documents on Slideshare is to use the Image Search features of search engines. Some user accounts on Slideshare represent organizations, and others represent individuals.

User profiles on Slideshare have these structure data fields:

- *hcard-fn, person-name, metatags-og_title,* and *metatags-slideshare_name* – the full name of the person is present in several places

- *hcard-title, hcard-role, person-role, person-jobtitle,* and *metatags_slideshare_work* – the user's job title

- *metatags-slideshare_organization* – the name of the organization where the individual works

- *person-location, postaladdress-addresscountry,* and *metatags-slideshare_location* – country

- *postaladdress-addressregion* – state, province, or other country sub-division

- *postaladdress-addresslocality* – city or town

- *person-description* – a truncated version of a user's profile's "about" section of the user profile

- *metatags_og_description* – significantly longer excerpt of the user's profile's "about" section (two to three times longer than the *person-description* field)

- *metatags-slideshare_following* and *metatags-slideshare_followers* – representing the number of people the user follows or is followed by (respectively) on Slideshare

- *metatags-slideshare_joined_on* – exact timestamp in coordinated universal time (UTC) when the user joined Slideshare

- *metatags-slideshare_updated_at* – exact timestamp in UTC when the user's profile was last updated

- *metatags-slideshare_upload_count* – the number of files the user has uploaded to Slideshare

Organizational profiles are similar but may contain the organization (*organization-name* and *organization-description* are the most common fields). Any single profile of either type may be missing one or more of these fields based on what profile fields the user in question has filled out.

Example searches using the Slideshare CSE:

- front end more:p:person-role:engineer – role contains *engineer*, *front* and *end* appear within the profile (front-end developers)

- pharmaceutical   more:p:person-role:sales   more:p:postaladdress-addresslocality:new*york – role contains *sales*, lives in *New York City*, profile mentions pharmaceutical (pharmaceutical salespeople in NYC)

- more:p:metatags-slideshare organization:kpmg more:p:postaladdress-addresslocality:paris more:p:postaladdress-addresscountry:france – works for KPMG and lives in Paris, France

The example Slideshare CSE also has three custom sorting options implemented: by the number of uploads, followers, and users followed.

REUTERS CSE OPERATORS

Reuters is an international news organization owned by Thomson Reuters. Until 2008, the Reuters news agency formed part of an independent company, Reuters Group plc, which was also a provider of financial market data. Since the acquisition of Reuters Group by the Thomson Corporation in 2008, the Reuters news agency has been a part of Thomson Reuters, making up the media division. It was established in 1851.

While the articles do not contain Schema.org objects, there are still interesting searchable fields among the metatags' data. Take, for example, the following:

- *metatags-og_title*

- *metatags-twitter_title*

- *metatags-analyticsattributes.contenttitle*

- *metatags-analyticsattributes.title*

- *metatags-sailthru.title*

Here, five different tags (respectively, one for Facebook's Open Graph, one for Twitter, and three for Reuter's internal analytics systems) contain the article's title! Information in Meta tags is often duplicated due to differences in standards. (When you share a link on Twitter, Twitter looks for one type of Meta tag to find the page's title and description, and Facebook looks for another type of Meta tag to find the information.) For the remaining examples, we omit duplicates, which add no new utility:

- *metatags-og_description* – This field contains (as does its omitted Twitter-oriented companion) a summary of the article's contents.

- *metatags-news_keywords* – This field contains a list of keywords to boost the articles ranking in searches for related terms.

- *metatags-author* – This field lists the author of an article.

- *metatags-og_article_section* – Many news feeds and news websites are organized by topical section, as print newspapers are, and this field represents the section to which the article belongs.

- *metatags-revision_date* – Nominally, this field contains the date of the article's most recent revision. However, in the majority of cases, it represents the publication date. The date is provided in the format *Thu Jun 11 20:59:16 UTC 2020.*

For examples of how we can use these fields in a search, take a look at the following uses of the Reuters CSE:

- more:p:metatags-news_keywords:semiconductors-more:p:metatags-author:reuters*editorial – This search looks for articles that include semiconductors as a keyword for search relevance boosting (and are therefore likely to be about a semiconductor company or the semiconductor industry), which have bylines other than Reuters Editorial.

- google more:p:metatags-revision_date:2006 – This search looks for articles about Google published (or updated) in 2006.

- vaccine more:p:metatags-news_keywords:uk more:p:metatags-og_article_section:business – A search to find articles in the business section about vaccines that have to do with the United Kingdom.

- manufacturing more:p:metatags-news_keywords:layoffs more:p:metatags-revision_date:2020 – Stories from 2020 having to do with layoffs and mentioning manufacturing would be found by this search.

- managingdirectormore:p:metatags-news_keywords:key*personnel*changes more:p:metatags-revision_date:2020 more:p:metatags-og_site_name:u.k. – This search finds articles from 2020 published for Reuters's United Kingdom audience that have to do with key personnel changes and refer to a managing director.

RocketReach CSE OPERATORS

RocketReach is a web-based tool that lets users find email addresses, phone, social links for over 250 million professionals across 6 million companies worldwide. It is a combination of tools that look for email addresses both from a company's domain and specific peoples' emails (*source:* crozdesk).

The RocketReach CSE helps to find email formats for companies.
Example searches:

- mount sinai hospital

- corporate recruiter

- "jane.doe"

- "jdoe"

- being used 100.0% of the time – This search will find companies that follow an email format 100% of the time. If you know someone's first and last name, and they work for such an employer, you can construct their emails.

DOXIMITY CSE OPERATORS

Doximity is an online networking service for medical professionals. Launched in 2011, the platform offers its members curated medical news, case collaboration, and messaging capabilities. The company was launched in March 2011 by co-founders Nate Gross, Jeff Tangney, and Shari Buck.

In 2018, the company announced that it had reached one million members, accounting for more than 70% of US physicians (*source:* Wikipedia).

Among specialized search sites, Doximity provides a remarkably rich structure to query. Here are some example searches using the Doximity CSE.

- more:p:person-honorificsuffix:md – MDs

- more:p:metatags-profile_gender:female – women

- more:p:educationalorganization-name:hopkins – school

- more:p:person-jobtitle:pediatric*nurse – job title

- more:p:person-jobtitle:surgeon more:p:postaladdress-addressregion:NJ – operator combination for job title and state (beyond *addressregion*, which refers to US states or comparable administrative regions in other countries, *addresslocality* refers to municipality, *streetaddress* refers to the street name and number, *addresscountry* refers to the country, and *postalcode* refers to the postal code).

XING CSE OPERATORS

XING is a Hamburg-based, career-oriented social networking site operated by New Work SE. The site primarily focuses on the German-speaking market, alongside XING Spain, and competes with the American platform LinkedIn.

OPEN Business Club AG was founded in August 2003 in Hamburg, Germany, by Lars Hinrichs. Its official debut was November 1, 2003. It was renamed XING in November 2006 (*source:* Wikipedia). XING has over 15 million members from Germany, over 1 million in Switzerland, and 1.5 million in Austria (*source:* XING).

XING public profiles have the following fields:

- *person-role* – this should be the person's job title, but on many profiles is actually the company a person works for due to an apparent error in XING's code

- *person-org* – this should be the person's company, but on many profiles is actually the person's job title due to an apparent error in XING's code

- *person-location* – location (free-form text)

- *metatags-og_title* contains the name, role, and company (e.g., *Tom Schillinger - Staff UX Designer - Mercedes Benz Research and Development North America*)

Keep in mind that many values are in German due to the nature of the Social Network.

Example searches using the XING CSE:

- more:p:person-org:SAP*CRM – job title contains *SAP* and *CRM*

- more:p:person-role:consulting – company name contains the word *consulting*

- more:p:person-org:managerin – women managers (German has a different word for men who are managers)

RESEARCHGATE CSE OPERATORS

ResearchGate is a professional network for scientists and researchers. Over 17 million members from all over the world use it to share, discover, and discuss research. They are guided by their mission to connect the world of science and make research open to all (*source:* https://www.researchgate.net/about).

ResearchGate was founded in 2008 by virologist Dr. Ijad Madisch, who remains the company's CEO, with physician Dr. Sören Hofmayer, and computer scientist Horst Fickenscher. It started in Boston, Massachusetts, and moved to Berlin, Germany, shortly afterward (*source:* Wikipedia). Examples:

- more:p:hcard-fn:jessica*brown

- more:p:person-role:investigator

- more:p:person-org:economy*research – employer name includes *economy* and *research*

- more:p:organization-name:Taylor*Francis – publisher name includes *Taylor* and *Francis*

- more:p:metatags-og_description:applied*mathematician

- more:p:organization-url:edu

- more:p:scholarlyarticle-headline:global*warming

- more:p:scholarlyarticle-datepublished:2019

Combinations:

- more:p:organization-name:university more:p:metatags-og_description: covid*19

- more:p:organization-name:house more:p:person-org:chemical

- more:p:organization-url:uk more:p:person-role:chemistry

GOOGLE SCHOLAR CSE OPERATORS

Google Scholar is a freely accessible web search engine that indexes the full text or metadata of scholarly literature across an array of publishing formats and disciplines. Released in beta in November 2004, the Google Scholar index includes most peer-reviewed online academic journals and books, conference papers, theses and dissertations, preprints, abstracts, technical reports, and other scholarly literature, including court opinions and patents. While Google does not publish the size of Google Scholar's database, scientometric researchers estimated it to contain roughly 389 million documents, such as articles, citations, and patents, making it the world's largest academic search engine in January 2018 (*source:* Wikipedia).

- more:p:hcard-fn:alan*scott

- more:p:hcard-title:Phd*student

- more:p:person-name:alexander*kirillov

- more:p:person-org:princeton – search for organization's email domain

- more:p:person-role:senior*scientist – (but person:role in most cases repeats the name of the employer)

- more:p:person-role:idaho*state*university

- more:p:person-title:professor

- more:p:person-jobtitle:postdoctoral*researcher

- more:p:person-worksfor:CIA,FBI – searching for employees of Central Intelligence Agency or Federal Bureau of Investigation

- more:p:person-worksfor:government

- more:p:person-worksfor:rockefeller

- more:p:itemlist-itemlistelement:lipid*nanoparticles

- more:p:itemlist-itemlistelement:small*molecules

Combinations:

- more:p:person-org:oxford more:p:itemlist-itemlistelement:mild*dementia

- more:p:person-org:nyu more:p:itemlist-itemlistelement:gender*equality

- more:p:person-worksfor:NHS more:p:itemlist-itemlistelement: immunodeficiency

- more:p:person-worksfor:NASA more:p:itemlist-itemlistelement:mars

- more:p:person-worksfor:FDA more:p:person-jobtitle:postdoctoral*fellow

VITALS CSE OPERATORS

Vitals claims to be "the largest online database of patient reviews for doctors and facilities." (*source:* Vitals.com). Founded by a trio from New Jersey in 2007 (*source:* EverybodyWiki), Vitals is now a part of WebMD (*source:* PRNewsWire). Unlike Doximity or LinkedIn, the Person object on Vitals profiles does not provide additional information about that doctor. Vitals' Person objects are about reviewers. To find information about doctors, we need to use the Physician instead.

Here are example searches using the Vitals CSE:

- more:p:physician-description:neurology

- more:p:physician-name:RN*MSN

- more:p:medicalspecialty-name:cardiovascular*disease

- more:p:aggregaterating-ratingvalue:4.5

- more:p:postaladdress-name:family*practice

- more:p:postaladdress-newpatient:accepting

- more:p:postaladdress-postalcode:99504

Combinations:

- more:p:medicalspecialty-name:dermatology more:p:postaladdress-addressregion:ks more:p:postaladdress-newpatient:accepting

- more:p:postaladdress-telephone:876 more:p:postaladdress-addressregion:az

Zocdoc DOCTORS CSE OPERATORS

Zocdoc provides online appointment scheduling software for doctors' offices, and a free search and scheduling interface for patients. Patients can search for doctors by language spoken, medical specialty, location, conditions treated, insurance accepted, appointment availability, rating, and reviews. Patients can also leave ratings and reviews for doctors on the site. Zocdoc offers coverage for over 2,000 cities across the US. While there are over one million providers listed on the site, according to an analysis performed in 2017 by Jake Bialer of the NYC Data Science Academy, only about 47,000 profiles represented actual physicians using the site. The remaining profiles were placeholder stubs.

Provider profiles on Zocdoc reside in the /doctor/path. There are two types of profiles. The one for oral health providers (Dentists and Oral Surgeons) contains the *Dentist* Schema.org object, and all other providers use the *Physician* object. Regardless of the object used, each profile contains the following interestingly structured data for search:

- *medicalspecialty-name* – the doctor's medical specialty

- *metatags-og_description* – the doctor's biographical summary

- *metatags-og_type* – either *dentist* or *doctor*

- *postaladdress-addresslocality* – the municipality where the doctor's practice resides

- *postaladdress-postalcode* – the postal code for the doctor's practice

- *postaladdress-addressregion* – the US state in which the doctor practices

- *postaladdress-streetaddress* – the street address of the doctor's practice

Note that not all provider profiles that use the *Physician* object represent fully licensed medical doctors. Chiropractors, psychologists, optometrists, and other types of health-care providers are listed on ZocDoc as well.

The *Dentist* and *Physician* objects have the same structure, so the data fields are the same, with only the object prefixes varying:

- *physician-identifier* or *dentist-identifier* – the provider's <u>National Provider Identifier</u> number

- *physician-name* or *dentist-name* – the provider's full name

- *physician-honorificsuffix* or *dentist-honorificsuffix* – the provider's honorific (e.g., DO, MD, DDS); while there is also an -honorificprefix field, the value is universally Dr.

- *physician-telephone* or *dentist-telephone* – the telephone number for the provider's practice

- *physician-memberof* or *dentist-memberof* – the business name of the health-care practice of the provider

Here are some example searches making use of the <u>Zocdoc Doctors</u> CSE:

- <u>more:p:physician-memberof</u>   <u>more:p:physician-honorificsuffix:DO</u> <u>more:p:medicalspecialty-name:internist</u>   <u>more:p:postaladdress-addressregion:CA</u> – This search will find Physician (not Dentist) profiles of internist osteopaths who reside in California and work as part of a larger medical practice.

- <u>more:p:medicalspecialty-name:oral*surgeon</u> <u>more:p:postaladdress-addresslocality:new*york</u> – This search finds profiles of oral surgeons in Los Angeles.

- <u>more:p:metatags-og_description:harvard more:p:metatags-og_description:</u> <u>trained,studied</u> – A search to find physicians who went to Harvard. This search operates on the assumption that the biographical summary (contained under "metatags-og_description") may reference where a doctor went to school, especially for graduates of prestigious programs. Common ways to describe this background are *Harvard-trained* or *studied at Harvard.*

- <u>more:p:metatags-og_description:cognitive*behavioral,CBT</u>   – Cognitive behavioral therapy (CBT) is a specific form of psychotherapeutic practice, so this search seeks to identify psychiatrists who provide this type of therapy.

- <u>more:p:dentist children's OR children OR pediatric</u> – Dentist profiles often lack the metatags-og_description field for some reason, but we can still search for general keywords in the profiles, such as in this search for pediatric dentists.

SPEAKERHUB CSE OPERATORS

<u>SpeakerHub</u> is a social media network dedicated to public speakers, trainers, and moderators. SpeakerHub is headquartered in Brussels, Belgium,

with the intent of connecting speakers to events all around the world. It was founded in 2016 (*source:* SpeakerHub site and Facebook page).

SpeakerHub CSE example searches:

- more:p:person-jobtitle:ceo

- more:p:person-affiliation:university*alabama

- more:p:person-jobtitle:researcher

- more:p:person-description:my*story

- more:p:metatags-profile_gender:female

- more:p:person-memberof:women

- more:p:organization-name:marketing

Combinations:

- more:p:person-jobtitle:chief more:p:person-affiliation:columbia

- more:p:organization-name:space more:p:person-jobtitle:director, manager

CLUSTRMAPS CSE OPERATORS

Clustrmaps.com combines information about addresses, people, and businesses from public records and online sources. It then presents the aggregated data about each entity into a combined profile. The site claims to have information about over 200 US persons, 23 million companies, and 107 million addresses. The available data varies by the type of record, but for persons, it typically includes an address and phone number and often an email address.

CSE Clustrmaps

Example searches (they are self-explanatory):

- more:p:person-location:Denton*MD

- more:p:person-gender:male

- more:p:person-telephone:760

- more:p:postaladdress-addresslocality:topanga
- more:p:postaladdress-postalcode:90290
- more:p:postaladdress-addressregion:CA
- more:p:postaladdress-addresscountry:US
- more:p:postaladdress-streetaddress:bryce*canyon*rd
- more:p:person-email:gmail
- more:p:person-email:org

Combinations:

- more:p:person-gender:female more:p:postaladdress-postalcode:20002 more:p:person-email:gmail
- more:p:person-location:portsmouth*NH more:p:person-email:yahoo
- more:p:person-email:walmart,ebay,coles,tesco,sears,walgreens more:p:person-location:NY

Summary

Congratulations on making it to the finish line! We hope you have enjoyed learning about the little-known techniques and creating CSEs.

Our book provides the most complete, accurate, and detailed coverage of CSEs and their practical use available on any media. Such a document has never been in existence; this is the first.

We have highlighted and shared examples of 15 features of CSEs:

1. (Invisibly for the user) include only given site(s) (e.g., *linkedin.com/in*, which points to LinkedIn profiles)

2. Exclude given site(s)

3. Give priority to the listed site(s) but search the entire web

4. Give priority to pages with given keyword(s)

5. Narrow to a language

6. Narrow to a country

7. Boost results by a country

8. Include sites via patterns using an Asterisk (e.g., *site:behance.net/*/resume*)

9. Automatically append a string to a user's search (for example, narrow the search to PDF documents by adding filetype:PDF)

10. Define synonyms to process user's input

11. Use the Synonyms feature to run long OR statements (for example, search for common women's names)

12. Select pages with given Schema.org object(s) (like *Person*, *Physician*, or *Organization*)

13. Search for the presence of Schema.org objects' fields (for example, find pages that have the filled "email" in the *Person* object)

14. Search within Schema.org objects' fields (for example, search for Github profiles containing "I love Python" in the bio or LinkedIn profiles containing "open to new opportunities" in the headline)

15. Guide the search by selecting Knowledge Graph object(s) (for example, find pages that are "CVs")

We hope you have enjoyed the book and are inspired to create, utilize, and share your Custom Search Engines as well as perform advanced searches using CSE Operators.

Discover *more:*! (We trust the book title is clear to you by now.)

Glossary

Annotations In the context of Custom Search Engines, annotations represent the Sites to search and Sites to exclude. When you have a large number of URLs or URL patterns to add to your CSE, Google recommends uploading an annotations file with the details instead of entering them in the Control Panel.

APIs Application Programming Interfaces are a set of instructions for third parties to access the internal functionality of a piece of software. Google offers several APIs for Custom Search. They allow for building third-party software that includes Custom Search Engine results. These APIs return results in an easy-to-parse format. The results include structured search metadata that is hidden from users of the CSE or Google web interfaces.

Boolean Logic In the context of a web search engine, Boolean logical operators refer to the three basic operations in Boolean logic: AND (conjunction), OR (disjunction), and NOT (negation). They take their name from mathematician George Boole, who originally defined this set of logical operations. AND indicates that results must include each of the terms, OR - at least one of the terms, and NOT requires that results not contain a given term. On Google, AND is implicit, OR is represented by the word OR (capitalized), and NOT is represented by the minus (-) preceding the negated term.

Boolean search Searches that use the Boolean logic AND, OR, and NOT. Also, colloquially, any search that includes advanced search operators.

CAPTCHA An acronym for Completely Automated Public Turing test to tell Computers and Humans Apart, as well as a pseudo-homonym for capture. CAPTCHAs are a type of challenge presented by web pages that allow users to submit information (including search queries) to a web server. They require you to click on a checkbox and identify crosswalks, buses, bicycles, or traffic lights. In the context of search engines, these challenges require the user to submit additional information, such as the color of a picture, the solution to a mathematical equation, or the words pictured in an image. Users failing to do so may have their access to the web page blocked.

Configuration Files Custom Search Engines

Control Panel The main user interface for customizing and editing Custom Search Engines.Custom Search Engines Now officially named *Programmable Search Engines* by Google. Provide their creator the ability to create customized search experiences for end-users.

Dashboard List of Custom Search Engines in your account. It links to the Control Panel and Search Page for each CSE.

Google Index Much like authors provide the index at the back of a book, for every term, Google has a list of (potentially) hundreds of billions of web pages containing that term. Whenever Google's crawlers visit and gather information about a web page, they update the index for every term contained in the page. They also update structured data, including Schema.org, Meta tags, and microformats, contained in the page.

HTML HyperText Markup Language is the standard markup language used for the web. A markup language is a standard for annotating text with machine-readable instructions. Hypertext (from the Greek for over or beyond) means text with references (hyperlinks, now typically just called links) to additional information. A simplified way to think about HTML is as the set of instructions that web browsers follow for displaying the contents of web pages.

Image Search As much as Google's search index is based on keywords contained in a given web page, Google's image search index is based on keywords contained in the metadata for images. Though hidden from users in most web browsers, the HTML that tells browsers to display an image will typically contain metadata that includes the name and a brief description of the picture. This text may be displayed instead of the image if there is an error in loading the image or the web browser does not support viewing images. Google also indexes images by size, file type, color scheme, and other features (such as whether the image is line-art or depicts a person's face) and provides a separate search interface for images. Custom Search Engines may also include the option to search for images.

Including CAPTCHAs discourages the use of software programs to access a web page automatically. By now, the original CAPTCHA implementation is outdated, and many sites (including Google) have implemented newer challenges like Re-CAPTCHA.

It is possible to provide additional information in an annotation's XML file upload that you cannot enter via the Control Panel. For example, you can add a score to a URL or URL pattern that will cause results from that URL or pattern to appear higher or lower in the search results for your CSE. You can also export the annotations in your CSE into an XML file called (by default) annotations.xml.

Javascript A computer programming language that has developed into a core component of the web. Javascript is the primary mechanism by which dynamic elements of web pages (i.e., any content on a web page that can change without fully refreshing) are implemented. Dynamic pages pose a challenge to search engines, as the contents of the page may be different between a search engine's and a user's visits.

Knowledge Graph In computing terms, a graph is a way of representing the relationships between things. Google's Knowledge Graph is a model of how different concepts are related. This information helps Google to differentiate between, for example, *Paris, France*, and *Paris, Texas*. The Knowledge Graph objects in Custom Search Engines allows you to search for categories of related terms, even if the keywords used to describe the individual terms are different. For example, including the Knowledge Graph object named *Resume* in your CSE search would show results that are conceptually related to Resumes, even if the actual keyword *Resume* or closely related terms like *CV* are not present within the results.

Meta Tags A type of HTML tags containing information that tells User Agents (web browsers and web crawling programs used by search engines like Google to build their indices) how to display and process the code that makes up the web page. There are many uses for Meta tags, such as adjusting the way text and images are laid out on a page depending on the size of the screen or window. Within this book, we are most interested in the application to search engines.

Meta tags, much like Schema.org objects, define data fields, which in turn contain values. Meta tags relevant to search typically contain an attribute called *name*, with the name of a value representing the data field, and an attribute called *content*, with the field value. In the same way we would search for *more:pagemap:person-name*, we can also search for *more:pagemap:metatags-description* to look for pages that contain a Meta tag with the name attribute containing the value *description*.

Microformats Predated Schema.org and were the first widespread effort to establish a markup standard for web page metadata. Many pages include both microformats and Schema.org. Only the hCard and hCalendar microformats have been formally standardized, though many others have been proposed as drafts. Custom Search Engines can reliably search for hCard and hCalendar.

Programmable Search Engine see *Custom Search Engine*

Refinements In a Custom Search Engine, refinements are special elements added to the end-user search interface. When clicked, additional search parameters (filters) are applied, further narrowing down the results.

Schema.org A community-built standard vocabulary for providing structured data about the contents of web pages. Major search engines "understand" objects defined by the Schema.org standard.

Search Operator A search query consists of one or more terms. Each term may be a keyword (e.g., *apple*) or search operator (e.g., *inurl:*, which tells Google to find pages with a word or phrase contained within their URLs). Search operators often take a keyword as an input (e.g., *inurl:profile*). Google's search operators follow the format of being all lowercase, using a colon (:) between the operator and the input. The exceptions to this rule are the OR Boolean operator, the ""s (phrase) operator, and the * (wildcard) operator.

SERP SERP is the abbreviation for Search Engine Result Page or one page of results from a search engine.

Snippet A brief preview of the text content of the web page. Typically, a result's snippet will include either specific text specified by the web page publisher or the text that Google finds most relevant to the search.

URL Patterns A URL pattern uses the asterisk (*) to explicitly define the format of acceptable (or unacceptable) results. So, **.example.com/** is a URL pattern that looks for results from *example.com* that include any subdomain in their URLs and have some additional path in their URLs. The URL pattern *example.com/*/*/profile* finds results from the domain *example.com* with the word *profile* at the end of the URL that have no subdomain and must include two directories in the path that precedes the term *profile*.

URLs *URL* (plural URLs) is an abbreviation of *Uniform Resource Locator*, the technical term for what is commonly called a "web address" or a "page address." Typical web page URLs (e.g., *https://www.example.com/directory/file.pdf*) contain the scheme (*https*), host (*www.example.com*), and path (*directory/file.pdf*). The host is further broken down in *subdomain(s)* (in this case, *www*), domain (*example*), and top-level domain (*com*). The path in a URL roughly corresponds to the directory and file structure used on a computer. In the context of the CSE Control Panel, full or partial URLs can be listed in the *Sites to include* and *Sites to exclude* settings. When including a partial URL, you can remove components from the left side of the ULR (https, *www*, even *example*), so *www.example.com*, *example.com*, or even just *com* are valid entries. All URLs added in the *Sites to include* or *Sites to exclude* settings must include at least a top-level domain, however. When you list a URL in the *Sites to include* or *Sites to exclude* settings of a CSE, any results with a URL that matches or includes the URLs (or partial URLs) will be included or excluded from the results.

A CSE that includes *www.example.com* will not find results from *ww2.example.com* or *users.example.com*. However, a CSE that includes *example.com* will also find results from any subdomains, and a CSE that simply includes *com* will find results from any *.com* website.

X-Ray An X-Ray search is a term used in the staffing and recruiting industry. Named after the medical procedure of taking an X-Ray image to see inside the human body, it means to use the *site:* search operator to search for results within a website. The *site:* operator limits search results to those from a given domain.

Appendix A

List of Our CSEs

The following list shows where you can find the first appearance of each CSE in the book. Where the CSE is not described before its first appearance, we have also indicated where to find the description.

Appendix B

Complete List of Google Search Operators

Operator	Meaning
	Pages containing keywords in:
allinurl: / inurl:	the URL
allintitle: / intitle:	page title
allintext: / intext:	page text
allinanchor: / inanchor:	the anchor text on the page
filetype:	file types
site:	narrow results to a site
related:	shows similar sites (the operator is being phased out)
info:	shows page info
define	gives a definition
The quotes ("")	search for a phrase
The minus (-)	word or phrase or site exclusion
OR	alternatives
Numrange (..)	search for a range of numbers
Asterisk (*)	stands for a word or a few words
AROUND (n)	proximity search
before:, after:	date search

About the Authors

Irina Shamaeva is a Partner and Chief Sourcer at Brain Gain Recruiting, Blogger at Boolean Strings, a world-class trainer providing webinars on Sourcing, conference speaker, and co-founder of SocialList.io. Her first book, 300 Best Boolean Strings, is now in its fifth edition.

Previously, Irina was a Software Engineer and Manager at several startup and biotech companies in the San Francisco Bay Area. She holds an MS in Mathematics from Moscow University.

David Galley is a self-described "Sourcing nerd" and enjoys digging deep into technical details, discovering information others may have missed. David heads up the development and delivery of the Sourcing Certifications Training Library and Certification Exams. He also runs custom online and on-site training programs for corporate teams and recruiting agencies. David holds a BS in Accounting from Colorado Technical University.

Index

Page numbers in *italic* indicate figures.

Printed in the United States
by Baker & Taylor Publisher Services